T0200286

The Universal History of Us

The Universal History of Us

*A 13.8 billion year tale from the
Big Bang to you*

TIM COULSON

MICHAEL  JOSEPH

PENGUIN MICHAEL JOSEPH

UK | USA | Canada | Ireland | Australia
India | New Zealand | South Africa

Penguin Michael Joseph is part of the Penguin Random House group of companies
whose addresses can be found at global.penguinrandomhouse.com

First published by Michael Joseph, 2024

001

Copyright © Tim Coulson, 2024

Set in 13.5/16pt Garamond MT Std
Typeset by Jouve (UK), Milton Keynes
Printed and bound in Great Britain by Clays Ltd, Elcograf S.p.A.

The authorized representative in the EEA is Penguin Random House Ireland,
Morrison Chambers, 32 Nassau Street, Dublin D02 YH68

A CIP catalogue record for this book is available from the British Library

HARDBACK ISBN: 978-0-241-66230-4
TRADE PAPERBACK ISBN: 978-1-405-96887-4

www.greenpenguin.co.uk

MIX
Paper | Supporting
responsible forestry
FSC
www.fsc.org   FSC® C018179

Penguin Random House is committed to a
sustainable future for our business, our readers
and our planet. This book is made from Forest
Stewardship Council® certified paper.

For Sonya, Sophie, Georgia and Luke

'Contemplate the marvel that is existence, and rejoice that you are able to do so. I feel I have the right to tell you this because, as I am inscribing these words, I am doing the same.'

– *Exhalation* by Ted Chiang

Contents

Introduction

If I could be a god there would be no flashy miracles, but there would be lots of universes. Enough universes to enable a scientist like me to conduct an ambitious experiment – an experiment that I, and I'm sure many others, dream of. I would like to identically re-create the conditions at the instant our universe came into being, run the clock forward, and examine how many of the resulting universes develop to become a home for you and me.

I would ensure that each starting point for each of my universes was identical. After 13.77 billion years, the age of our own universe, I would head to where our solar system should be in each of my experimental replicates. If our solar system was there, I would search out Earth before looking for you and me. Would we always be there, just as we are today? Or would we only be there in a few of the universes? Or in none at all? Would little green aliens with giant ears and no noses sometimes be in our place? On some occasions perhaps there would be no sign of our sun or the Earth at all, or maybe the Earth would be barren, moonless and devoid of plants and animals, circulating too far from the sun for life to flourish. In some universes life may be abundant, in others it might be rare, or even entirely absent. As far as we know, we are the only planet with life in our universe, although we have only explored a very tiny part of it.

My experiment would reveal whether the history of our universe was set at its birth or whether each replay produces a different outcome. If each play of the tape produced you and me, scientists would describe the universe as being deterministic. With hard work and a huge amount of computing power, we could perfectly predict the future of any universe from its birth until its end. Some physicists have argued that this is what would happen if the experiment I describe was run. They believe that physics will one day reveal that things we currently think are random, including the behaviour of tiny particles, will be shown to be deterministic. In contrast, if each outcome was different, as most biologists and many other scientists believe, then that means random events have influenced the development of our universe. A random difference in the history of each universe could result in very different outcomes. Perhaps if the asteroid that killed the dinosaurs had passed by Earth rather than hitting it, humans might never have evolved and an intelligent species of dinosaur might be writing a book like this instead of me.

Scientists describe universes where chance events make it impossible to accurately predict the future as being stochastic. If random events do play a role in the history of our universe and our existence, we will not be able to perfectly predict the future, but this does not mean there is no predictability. If we were to monitor 1,000 universes in my rerun experiment, perhaps intelligent life would arise 531 times but fail to evolve on 469 occasions. If you placed a wager on intelligent life evolving at the Betting Shop at the Beginning of Time, you would have just over a 50 per cent chance of winning that bet. With odds of a little less than two-to-one, if intelligent life did eventually evolve, you would win twice as much as you originally bet.

Sadly, I am not a god, and we do not have the know-how to run my thought experiment to assess whether the universe is deterministic or stochastic. Creating new universes on a whim, and studying their evolution, is beyond our technical expertise. Therefore, answering the question of whether you and I were inevitable at the birth of our universe, or if we are just incredibly lucky, requires another approach: science that we can do. This book aims to tell two stories in parallel. The story of the universe from its birth until you and me; and the story of what had to happen for you and me to exist. These stories are a 13.77-billion-year epic. They involve unimaginable violence, death and a lot of sex, and although they often read as a tragedy they are ultimately a story of success. The sleuths who have slaved to piece together the stories of why you and I are alive (and know it) are more imaginative and cunning than Miss Marple, Sherlock Holmes, Nancy Drew and Hercule Poirot combined. They count among their number Albert Einstein, Marie Curie, Isaac Newton, Rosalind Franklin and Charles Darwin, and thousands of others of whom you will likely have never heard. The story is a work in progress, and the plot is subject to edits as brilliant minds continue to solve more of the mystery of why I am me, and you are you. However, even the incomplete script is sufficiently awe-inspiring to be a story that everyone should know.

I find it humbling that scientists have been able to piece together so much of the story from the Big Bang to you and me. Perhaps even more humbling is earning my living as a scientist who conducts research to identify tiny new bits of this story, so that I can add this new understanding to the remarkable insights so many other scientists have made. I am so impressed by this story that I decided I wanted to try to tell it in a way that makes it accessible to everybody, and not just other scientists. I want to do this because I have met so many people who are unaware of the remarkable progress of science and, because of this unawareness, are distrusting of it. Science can be difficult and intimidating, the language used is often technical, and the concepts can be hard to grasp. Although I personally enjoy reading texts on the chemistry of marine hydrothermal vents, or models of the first microseconds of our universe's existence, I frequently have to look up the meaning of words despite three decades of working as a professional biologist. In this book, I want to explain things in words that the reader doesn't have to look up.

Humans are mind-bogglingly complicated. If you or I were to be dismantled, bit by bit, with each type of bit placed in its own pile, the 30 or more trillion cells (the building blocks of all life) that your body consists of would form roughly 220 piles. One pile would consist of red blood cells, another of nerve cells, another of skin cells, and so on. Thirty trillion is a vast number, the first of many vast numbers that we will encounter in this book. It is nearly impossible to comprehend large numbers such as this, but we should at least try. Thirty trillion seconds is equal to 950,000 years, or 9,500 centuries. That amount of time is hard to imagine. We can comprehend a single second easily enough, but imagining 950 millennia in seconds is impossible. Yet 30 trillion is a small number compared with the number we get when we start breaking each cell down into its constituent parts: molecules. On average, each one of your 30 trillion cells contains over 40 million protein molecules that are key to you staying alive. Thirty trillion times forty million is 1.2 billion trillion (twenty zeros). Compared to other types of molecules in your cells, proteins are actually quite rare. The most common molecule in your body is water, with 99 per cent of all molecules and about 60 per cent of your weight (water is lighter than most of the types of molecules in your body) being H_2O. The number of molecules in your body gets much larger still when we start to count all of them, and not just proteins. If we were to next dismantle every cell in each of the piles of cells and create new piles of molecules, we would end up with well over 30,000 separate heaps. Some of these heaps would contain vast numbers of molecules, others many fewer. For example, you have only about 0.003 g of cobalt in your body, yet it is essential to keep you healthy. Most cobalt atoms are found in molecules of B_{12}, a vitamin that when deficient can lead to neurological problems, joint pain, blurred vision, and even depression.

We can continue our classification exercise further by breaking up molecules into their constituent parts, atoms, and creating a new set of piles. We would now end up with sixty piles. There would be a pile of carbon atoms, one of iron atoms, another of oxygen atoms, and so on. Next, we break up each atom in each of these piles into its component parts and create a new set of piles, and we end up with a pile of protons, a pile of neutrons and a pile of electrons – the names of the building blocks of atoms. Next, we can break the protons and neutrons into things called up quarks and down quarks, but we cannot divide these strange beasts, nor the electrons, any further.

Early in the universe's history, these particles emerged from energy, so a final step could be to transform these piles of fundamental particles back into energy. If we were to do that, we would end up with a lot of energy from which all things, including us, developed. The universe started as a microscopic pinprick of intense energy. How did it get from there to you and me? I will describe the steps that had to happen. In summary, and avoiding too many spoilers, some of the energy morphed into the quarks and electrons. When quarks interact, they become more complicated particles called protons and neutrons, which themselves can interact to form even more complicated things called atomic nuclei. These atomic nuclei then interact with electrons to make atoms, which in turn can interact to form molecules. There is a vast number of different types of molecules in the universe, and way more than the 30,000 different types that make up you. Under some circumstances these molecules can interact to form planets, cells and living organisms. Different species of these organisms interact in a range of ways, and over billions of years these interactions between species resulted in some of these organisms evolving to become more and more complicated until, eventually, humans evolved.

Scientists have uncovered our understanding of the universe through observations and experiments. A well-designed experiment is a great way to test an idea or a hypothesis, and to find out new facts. Many people find science daunting, and yet we are all scientific experimentalists at heart. You might not believe me, so I will elaborate. If you have a practical problem to solve, you probably experiment with several different approaches before finding one that works. Do you bribe a child to tidy her bedroom, sternly instruct her to do so, withhold pocket money until it is done, or lead by example? None of these approaches worked with my children, but the application of trial and error brought me more success with cooking. By experimenting with the quantities of different ingredients, and by tweaking the oven temperature and cooking time, I learned to produce the perfect British snack, a scotch egg.

Scientists experiment much in the same way as I do in the kitchen. They tweak conditions in the lab to see how the result is affected. The scientific outcome won't be a tidy bedroom or a tasty meal, but rather some insight into the way the physical, chemical or biological world works. Some experiments that scientists have conducted are monumental in proportion. The world's biggest machine is called the Large Hadron Collider, or LHC. It is located at the Conseil Européen pour la Recherche Nucléaire (CERN), and it consists of a 27-kilometre circular tunnel under Switzerland. Huge electrically powered magnets inside the tunnel accelerate particles to a hair's breadth of the speed of light, the universe's upper speed limit, before smashing the particles together. Vast detectors in underground cathedral-sized caverns record the outcomes of these collisions. Analysis of the information collected has revealed subatomic particles that the universe is constructed from, and these discoveries have led to some physicists claiming we are on the verge of constructing a 'theory of everything' that describes all the interactions of all the particles in our universe. We do not yet have a theory of everything, but physicists, chemists, biologists, mathematicians, historians, archaeologists and researchers in numerous other fields have made astonishing progress.

These inquisitive and clever researchers have revealed that there are many key events that were required for you and me to exist. They have made progress on identifying these events by asking carefully thought-through questions to explain why they occurred. We know that life exists in our universe by looking in the mirror, but could it have come about if gravity was just a little bit weaker, or if ice sank in water rather than floated? Volcanoes are thought to have played a key role in the birth of life, and Earth is unusually volcanic compared to other rocky planets we have studied. Tides may also have been necessary for life, and for those we have the moon to thank. Geologists think the moon was formed when a planet called Theia crashed into the young Earth. If Theia had passed by Earth, then we would have no moon and quite possibly no life. What if coal had not formed 300 million years ago in the carboniferous geological epoch? Would we have worked out how to use sunlight and wind to power our technological development and would we have avoided anthropogenic climate change? Our technological supremacy is almost certainly dependent on the 300-million-year-old dead plants from which coal is formed. Does this mean coal must form on all planets that harbour technologically advanced civilizations and, if it is a requirement, how often does it form? As we have made progress in answering some of these questions we have been able to tackle other important questions too, such as how common might life be in our universe: are civilizations like ours two a penny, or is ours unique?

The history of the scientific stories that have provided insight into the first 13.77 billion years of our universe is full of examples of perseverance and remarkable and sometimes surprising discoveries. There were many scientific dead ends, when good ideas proved to be wrong, but also many eureka moments when sense was made of bewildering observations. Many of these scientific breakthroughs led to technological developments that have defined the modern world. GPS, solar panels, nuclear power, high crop yields, modern medicine and smartphones are only possible because of the insights hard won by scientists. A few of these researchers achieved celebrity, and another few became rich

The Universe Through Time

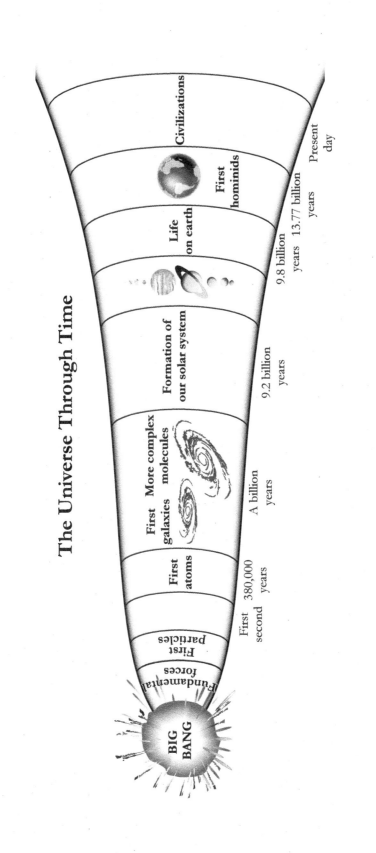

BIG BANG

Fundamental forces

First particles

First second

First atoms

380,000 years

First galaxies

More complex molecules

A billion years

Formation of our solar system

9.2 billion years

Life on earth

9.8 billion years

First hominids

13.77 billion years

Civilizations

Present day

from their discoveries, but the vast majority of advances were made by dedicated individuals who were driven by a desire to solve problems and to understand some aspect of the universe, rather than by a desire for fame and fortune. Humanity owes an enormous debt of gratitude to countless researchers who have spent careers trying to understand natural phenomena and the history of our universe. I largely avoid describing the stories of how scientists discovered the science I describe, not because they are uninteresting but rather because I want to focus on what we know. To also cover how we found out what we know would result in a much longer book.

Describing 13.77 billion years of history in a single, short book required making many decisions regarding what to include and what to exclude. I focused on the key things that had to happen for you and me to exist. The universe had to form, and four fundamental forces (gravity, electromagnetism and the strong and weak nuclear forces) had to emerge and have the right strengths. Quarks and electrons needed to come into existence before the quarks combined to produce protons and neutrons that in turn joined with electrons to create atoms of hydrogen and helium. The first stars needed to form to create heavier elements such as carbon and oxygen, and the elements then had to interact to produce a vast array of molecules. The Milky Way, solar systems and in particular the sun and the Earth had to form, and conditions needed to be right to kickstart life. Conditions on our planet needed to stay within bounds for life as we know it to spread and evolve, eventually leading to complex, conscious animals like you and me. Various events then moulded our personalities. I describe the science of these key events before discussing whether they were inevitable, or if we were simply very lucky.

What is included could be described in much more detail, with the scientific research behind the insights described in each chapter filling entire libraries. My aim is to show what astonishing progress researchers have made in understanding so many aspects of our universe rather than to describe nuances and ongoing research on specific details. If you want to learn more, I provide a list of a few science books in an appendix that go into much more detail about physics, chemistry, earth science, evolution and human history.

You may wonder what background and expertise I have. Unless you are a researcher in the fields of ecology and evolution, there is very little chance you will have come across me. This is my first foray into popular science writing, so I have no catalogue of previous titles you could have read. I am a professor at the University of Oxford and am currently joint head of the Department of Biology. I have been working as a scientist in research institutes and universities for over thirty years, and my research focuses on understanding how the natural world works. When I teach, I strive hard to make complex ideas accessible, and I also get a kick out of giving public lectures.

When my wife, Sonya, told her close friends she was dating an Oxford don shortly after we got together in 2013, by their own admission they assumed I'd be aloof, pompous and arrogant. Not because they had met other Oxford professors but because scientists, particularly those employed by leading universities, don't always have an endearing public image. I was pleased her friends liked me, and that I did not match their expectations. I have met aloof, pompous and arrogant Oxford dons, but I have also met lawyers, accountants and titans of industry with similar characteristics. Although science can be a daunting subject, scientists are just like everyone else. Some are humble, others not. They can be anxious, make mistakes, be serious or funny. I never chose to be an Oxford professor, and it instead happened through a series of fortunate and unplanned events, some of which I describe in this book as I interweave bits of my personal journey into the narrative. I do this for a number of different reasons. First, I not only want to make science accessible, I also want to reveal some of the human side of being a scientist. Second, towards the end of the book I explore how personalities arise. As a bit of this chapter focuses on how key events in my life may have shaped me, it is necessary for you to know a little bit about the sort of person I am. Finally, in the closing chapter, I touch upon what I have learned about what science has taught us and, once again, a little bit of my history helps the narrative.

Writing this book is part of a personal journey that started over thirty years ago when I was in my early twenties. As a teenager, I did not know what I wanted to do when I grew up, but I applied for and was offered a place to study maths at a university in the UK. I figured that being trained in mathematics to a high standard would keep my employment options open, and I was lucky in that I enjoyed the subject. Before starting university, I spent a year teaching in a rural school in Zimbabwe, and on one walk through the bush I stopped to watch a herd of antelope. I decided that being a biologist might be a better fit for me than being a maths graduate, and so I switched courses and ended up studying biology at the University of York in the north of England.

During my year of teaching, I fell in love with Zimbabwe and with the people I met there. I yearned to return to Africa, and the opportunity arose when it came to choosing an undergraduate research project that would be conducted at the end of my second year. It turned out it was possible to design your own project if you could persuade a faculty member to supervise it. I was lucky in that I managed to do just that.

A well-connected cousin was kindly able to arrange for me to go and stay at Kora National Reserve in Kenya, in the bush camp of George Adamson, an old-school conservationist who had shot to fame following the release of the book and film *Born Free*, which covered parts of his life and the life of his wife, Joy. It was Joy who wrote the book. My project compared the behaviour of wild lion cubs with hand-reared ones; their mother had been shot and George was raising them for release back into the wild. I spent a few weeks at the camp, watching and recording lion behaviour, before my then girlfriend and a couple of other friends flew to Kenya and we hitchhiked to Tanzania, and then Malawi, before flying home.

My time in Africa was once again fabulous, and I enjoyed every minute of it, even if my research project was scientifically underpowered. As a small child, I'd wanted to be Tarzan when I grew up. As a 21-year-old man working with lions, I got as close to that dream as I ever would, but I wasn't very good at it. Nonetheless, there were various important African experiences that set me on the path to becoming a scientist.

At some point during my travels after leaving Kenya, I contracted cerebral malaria, caused by the pathogen *Plasmodium falciparum*. I had been taking prophylactics, but resistance to the drugs I was on was starting to emerge. The fact I had muddled up my water-purification and antimalarial drugs did not help. The water I was drinking was not potable, and I developed a waterborne stomach infection. While sick, I did not absorb the antimalarial compounds, and I caught the disease. It also explained why every time I took what I thought were my malaria pills I had a pain in my guts; it was the chlorine in the water-purification tablets burning the lining of my stomach.

Cerebral malaria comes in waves every few days, and I had been unwell in Africa but hadn't known why. A day after my return to the UK, I became delirious and was rushed to Addenbrooke's Hospital near my parents' home, just outside Cambridge. I don't remember much of this time, but I do remember that after the diagnosis and the beginning of treatment a doctor told me that the next episode would have killed me. Fortunately, this form of malaria can be cured, and in time I made a full recovery. I realized that I was lucky to have had this attack following my return, rather than by the side of the road while hitchhiking through Malawi, and I found this thought sobering. I had never before considered my mortality, and my brush with malaria went on to define aspects of my life.

Over the coming months I started to think about what I would like to achieve in my life. I thought about how I would want to feel on my actual deathbed. If I had died from malaria, I would have been disappointed with how I had wasted my earlier years. I decided that when I did die, I would like my last act to be to look back on my life and think it had been fun and that I had also achieved something. It was around this time that I also decided I did not believe in a god and started to truly embrace science as the way to find out new knowledge. I decided that by the end of my life I wanted to have a good understanding of why I existed, and why I had developed the personality that would soon blink out. That was also the period of my life when I decided I wanted to be a scientist.

I have spent the intervening years thinking about my existence and researching it. Many scientists want to make a difference, often by working on ways to reduce human or animal suffering, improve standards of living or address threats to humanity such as climate change. My motivation was more personal. I simply wanted to understand why I existed and what had to happen for me to be here.

I made these decisions over thirty years ago. It took me so long to get my act together because life intervened. I needed to earn a living, I married, had a family, divorced and married again. My life has sometimes been fun, sometimes sad, easy, hard, frustrating and rewarding, just like everyone else's. The desire to understand why I exist has remained with me to this day, and seeing other authors like Bill Bryson and David Christian write fabulous books on the universe made me think I might also be able to write a book, but one that takes a different approach to those that have gone before. I do not have all the answers as to why we exist, but I do have a good understanding of what had to happen, and I also have an appreciation of how scientists have worked out why each of the key events occurred.

As I worked as a scientist, I also came to appreciate that quite a lot of people were distrusting of science and tended not to engage with it. Science can be difficult, some results are unintuitive, and it has not always been taught in an engaging manner in school curricula. Some distrust in science also arises because the technological advances it has permitted in recent decades are astonishing, yet not all technology has been used to benefit humanity. Splitting the atom was a remarkable scientific achievement, yet it ushered in the age of nuclear bombs that have the potential to wipe out humanity. Lasers can be used in eye surgery but also as weapons to burn through flesh, while unmanned drones can be used to plant seeds from the air without churning up soil but can also be used to blow people up hundreds of miles distant from the drone operators.

Science is not to blame for the way it is used. When it is used to harm or kill it is because someone has chosen to apply science for harm. Similarly, when it is used for good, it is down to human choice. Science is just a way of finding out facts about the world. Positive benefits of science include extending human life by eradicating, preventing and treating numerous diseases, through to increasing food security across many parts of the globe. Whether such positive impacts have also increased human happiness is less clear. Technology buys us more time, both in making our day-to-day lives easier and in extending the number of years we can expect to be alive, but it can also generate anxiety. Could artificial intelligence beat me to writing a book like this? Will future versions of ChatGPT make me redundant? I don't know, but if people understand science, they can contribute to debates about how it should be used. As I focused on why we exist, I realized I also wanted to be an advocate for science. It is a harder subject than history or philosophy, but it does generate progress in a way that no other subject does. Without science and technology, we'd be sitting outside at the mercies of the weather, arguing over why events happened and speculating on whether earthquakes were due to us upsetting a powerful ancestral spirit or an omen of future strife. We now know, thanks to science, that earthquakes occur when huge slabs of rock that form our planet's surface slide over one another. Science is remarkable. It is why we have a good understanding of how we came to exist, but our application of it has also led to us significantly changing the world in recent decades. The next chapter is about how science works, and how we find out facts. We would know nothing about how the universe works and came to be as it is without the scientific method, the development of which is arguably humanity's greatest achievement.

The Scientific Method

Thales of Miletus has a claim to be the world's first scientist. He was also the first of the seven sages of ancient Greece, men famed for their achievements in philosophy, law and politics who were revered for their wisdom and knowledge. Thales lived 2,600 years ago in the city that takes his name and, if everything written about him is true, he led a remarkable life and was quite a celebrity.

Piecing together the life of Thales of Miletus is not easy, as many of the achievements attributed to him were recorded decades after his death. Accounts of his accomplishments from the time of his life are scarce, and historians have failed to find any texts he wrote. But he was held in high regard by ancient Greek historians such as Herodotus and Eudemus, who list some of his most remarkable achievements as measuring the diameters of the sun and the moon, working out that the year is 365 days long, improving naval navigation by using the stars, predicting solar eclipses, setting the summer and winter solstices, diverting the flow of the River Halys so an army could cross it, developing numerous mathematical theorems, inventing economic futures trading and determining that the Earth is spherical. Evidence for some of these deeds is more compelling than for others, but one universal theme runs through these histories: Thales sought understanding via hypotheses he could test.

Scientists, from Thales of Miletus to the modern day, strive to discover facts about the world. They do this using something called the scientific method. It is not complicated, but it is incredibly powerful, and it is the way that all the facts described in this book have been identified.

Science is about explaining why a particular observation happens and then predicting future occurrences. The simplest observations are of things that always happen, such as jumping into the air and falling back to Earth. When we jump up from the ground, we never float off into outer space, and we have gravity to thank for this. Gravity always pulls us down, but other observations are more fickle, sometimes occurring and other times not. Ice, when left alone, will sometimes become liquid water, and other times will remain as ice. When the temperature is below freezing (when you're at sea level on Earth, at least), ice remains ice, but when it is above freezing, ice will become liquid water. Science aims to make sense of observations, and to identify the circumstances when particular ones occur. Jumping into the air, and ice melting or not, are relatively simple observations that we can endlessly repeat.

Rarer events are harder to study. Observing that we exist and working out whether you and I were inevitable at the universe's birth is a much harder observation to understand and predict because it has only happened once, Nonetheless, scientists have been chipping away at the problem. Events that happened long ago are also difficult to study. The further back we look in time, the less information is typically available. We know much more about the life of modern celebrities, and often way more than we might like, than we know about eminent people in ancient Greece. Yet we know more about the inhabitants of ancient Miletus than we do about those of Çatalhoyuk, the world's first city, which reached its zenith about 9,000 years ago. Knowledge gets even scarcer when we go even further back in time. The fossil record that contributes to our understanding of the history of life is sparse, and usually patchy. Very few ancient individuals of any species became fossils after death. To date, palaeontologists have unearthed thirty-two adult *Tyrannosaurus rex* fossils. *T. rex* were large animals with massive bones and teeth, the parts of animals that are most likely to become fossils. You might think that thirty-two sounds like quite a lot, and compared to some species it is, but this fearsome and massive dinosaur species roamed the Earth for 2.5 million years, and palaeontologists have estimated that a total of 2.5 billion *T. rex* individuals may have lived. If these rates of fossilization translated to humans alive today, the remains of only about 100 of us will survive as fossils 66 million years hence. That translates to 0.00000128 per cent of people becoming a fossil.

For an animal (or plant) to become a fossil, it must die under the right circumstances in the right environment. If you want to maximize the chance of becoming fossilized, your best bet is to be rapidly buried in sediment from a flash flood or ash from a volcanic eruption, and for your corpse to lie undisturbed for at least 10,000 years, yet this is still no guarantee. I would like to become a fossil – my children argue I already have – although I'd rather have a more peaceful end than being buried alive by flood or eruption. I'd like to become a fossil because it would help future scientists interpret the history we are creating today. If future palaeontologists were to discover my fossil, I may prove more useful to humanity after death than during my life.

The history of organisms alive in the past is partially written in impressions in rocks, but these etchings are rare and are often little help when working out part of the story of how you and I came to be. Despite this, palaeontologists studying fossils have uncovered remarkable insights about the history of life.

In this chapter I want to explore how science works. The scientific method consists of several techniques that allow us to find out things about the universe in which we live. Contrary to the belief of some populist politicians and social media gurus, repeatedly saying fictitious stuff over and over again is not how knowledge is created. It is a very outdated way of trying to make sense of the world. Prior to knowledge gained via the scientific method, people would explain observations of the world around them with stories. Many of these are great yarns, but they are often nothing more than myths.

Given how much knowledge the scientific endeavour has created, I am surprised and disappointed at the high-profile role that myths and myth-creation continue to play in today's society. I have met conspiracy theorists who believe that the royal family is a race of alien lizards, that Elvis is still alive and in hiding, that Donald Trump won the 2020 US election, that the moon landing was faked, and that Keanu Reeves is immortal and may in fact be a vampire. Some of the more nonsensical theories may be laughable, but when people embark upon a path of rejecting facts for unsupported nonsense, they can do harm. Science has convincingly shown that neither homeopathy nor distance healing works, and that ginger cannot cure cancer. Yet many people cling to these misguided beliefs. Science and medicine cannot cure all diseases, but they have made astonishing progress in treating many. Turning one's back on modern medical treatments because a mystic you have never met claims to be praying for you in Timbuktu risks shortening your life. Trust in science.

It produces evidence to support or refute hypotheses and to explain the unexplained. Myths and conspiracies are not based on evidence. Our civilization is advanced because of science, not because illuminati are acting as master puppeteers, pulling strings from behind the scenes. The facts of why you came to exist are more inspiring than any myth or conspiracy theory can ever be, and they can also be verified.

The scientific method starts with an observation, often posed as a question. The observation could be about any aspect of the natural, or human-built, world. It might be as simple as 'why is that tree there?' or 'why is it that sort of tree and not another type?'; or 'why do I get ill?' and 'why do I get better?'; or 'why do I always return to the ground when I jump in the air?'; or 'why does it not snow as much in England now as it did when my parents were young?' The observation I address in this book is 'you and I exist', and the

The Scientific Method

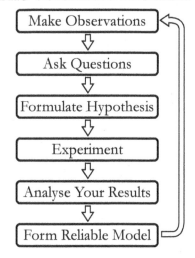

23

question I ask is 'why are we alive?' Science has taken us a long way towards answering this question.

I have asked questions like this from a young age and suspect they sometimes maddened my parents. The first time they took me to see the sea at the age of three or four, I apparently stared at the waves for several minutes before asking why they went up and down. My parents are well educated, and although they both have a grasp of science I was soon asking questions they could not answer, so we borrowed science books from the library. By the time I was a teenager they had bought me a subscription to a weekly science magazine for young adults. I looked forward to its arrival and would read it from cover to cover, sometimes drawing graphs or trying to write equations to help me understand some of the concepts I had read about.

If a psychologist had studied me as a child, they would have classified me as an odd, geeky kid. If they had studied my parents, they would have observed that my science fixation, coupled with a poor performance at school, sometimes gave my parents cause for concern. I didn't like school, not only because most of my cohort thought I was a bit of a weirdo – dressing in golfing trousers, pointed leather shoes, a trench coat and a green trilby for much of my teenage years didn't help my cause – but also because we were not taught how the world worked but were rather given lists of facts to learn. I was never taught the scientific method at school. I researched it myself at home.

After making observations, the next step in the scientific method is to pose a hypothesis. A hypothesis is a plausible explanation of a certain observation. A psychologist observing me as a kid might hypothesize that I was genetically predetermined to be a science geek, but back in the early

1980s they could not have easily tested this. Hypotheses are only useful when they can be tested with further observations, with experiments, or with both. An untestable hypothesis is equivalent to a myth or a story: it may sound compelling, but knowledge cannot extend beyond what we can observe.

To demonstrate how to pose a hypothesis I will focus on the first of the questions I asked above: 'why is that tree there?' I chose this question for two reasons. First, it is something we can all relate to, but it may be a question you may have never asked while gazing at a tree. Second, just after I started studying for my Ph.D., I sat gazing at a silver birch in the middle of a field for an hour or so, pondering this exact question. It motivated the topic of my doctoral thesis: how do animals like squirrels and deer contribute to the distribution of trees across a landscape?

The immediate answer to the question of why a tree is in a particular location is that a viable seed of a tree species arrived at that location, that the conditions were right for the seed to germinate, for the seedling to flourish and survive to become a sapling, and for the sapling to grow to become an adult tree. To do this it needed to survive the threat of death from herbivores devouring it, or from pathogens infecting it with a fatal illness, while acquiring enough water, light and nutrients to grow. Although this answer is obvious, what hypotheses need to have been posed and tested to acquire this knowledge? It turns out, quite a few.

It must first be hypothesized and tested that trees develop from seeds, something that is now obvious to you and me. The hypothesis can be tested via an experiment that would involve collecting seeds produced by several species of tree, planting them in known locations and seeing what they develop into. Given our knowledge today, you might think this hypothesis is ridiculously trivial. But as recently as the mid nineteenth century many people still believed in the theory of spontaneous generation, where living beings could arise from non-living objects such as rocks and water. They held this belief jointly with the knowledge that crops grew from seeds, but the necessity of a seed to produce a new seedling was not universally accepted wisdom as life forms were thought to spontaneously arise under some circumstances. A few writers in the seventeenth century, for example, argued that wheat and rags would spontaneously generate mice.

When seeds are collected and planted in an experiment, not all of them will germinate and grow into seedlings, but if enough do, we will have evidence that seeds can develop into young trees. Yet this insight naturally leads to a new question: why do some seeds fail to grow into trees? We might hypothesize that local conditions determine whether seeds germinate and survive to seedlings. An experiment could then be devised to plant seeds in different soil types, and to provision them with different amounts of light, water and nutrients to see which thrive. If seeds germinate in some conditions, but not in others, this provides evidence that environmental conditions can help explain the observation that not all seeds succeed in germinating into seedlings.

Even when conditions are suitable for seeds to germinate and thrive, not all seedlings will survive to become saplings. The next question might be 'why?' Some seedlings may appear to have suffered herbivory from animals like insects, snails, rabbits or deer. Others may look diseased, perhaps having been attacked by viruses, bacteria or fungi. These observations naturally lead to new hypotheses that disease and herbivory are important sources of mortality for seedlings. The next experiment is to compare the performance of seeds, seedlings and saplings grown in favourable light, water and nutrient conditions, but with some exposed to herbivores and pathogens, while others are protected from them. If death rates differ between these two groups, then this provides evidence that herbivores and pathogens play a role in determining whether seeds survive to grow into adult trees.

Even when water, nutrients and light availability are provided in just the right amounts, and pathogens and herbivores are absent, there may still be some seeds that fail to germinate. We could now hypothesize that there might be something wrong with some seeds. Perhaps they have not developed correctly, have gone off, or there is something amiss with their genes, which means they are unable to germinate. To test these hypotheses, we would need to move from the field and greenhouse to the laboratory. We might use microscopes, weighing scales, mass spectrometers, gene-sequencing machines and even genetic engineering to compare the morphology, metabolism and genetic code of seeds before planting them in ideal conditions and then monitoring which germinate and which do not.

For part of my Ph.D. I conducted experiments where I excluded deer from parts of an English woodland to examine how they impacted seed and seedling survival. I did this by fencing off areas that I termed 'deer-exclusion treatments'. I then chose unfenced areas of the same size that deer could access, and these were called controls. I went on to plant seeds and seedlings in each plot. I wanted the deer-exclusion treatment and control regions to be as similar as possible in all aspects except for whether deer could access them or not.

I cut back the forest understorey in each treatment and control area with a scythe. At one point, I got a little bored, and to relieve the boredom I waved the scythe around while stating, 'I am the Grim Reaper.' At that exact moment a woman walking her dog came around the corner, saw me mucking around and quickly walked away. My field site was close to a secure hospital for the criminally insane, so I rapidly left in case the dog walker reported to the police that a patient may have escaped. At least by then I had ditched the trench coat and green trilby. Despite the hiccough, the experiment worked, and I went on to estimate how deer significantly impacted seed and seedling survival of several species in my study site.

At the end of my field experiments, I had helped understand where and why trees might successfully establish themselves. That was a useful advance, but it did not address the second of the questions asked earlier in the chapter: why is the tree upon which you are gazing of one species and not of another? Why was the tree I gazed at all those years ago a silver birch and not a coco-de-mer palm? To answer this question, we need to generate a new set of hypotheses.

These hypotheses might revolve around whether particular conditions favour seeds from some species of tree over others, or whether the proximity of potential parent trees, and luck, result in a seed of a particular species arriving in a particular location at a time that happened to be appropriate for it to germinate and thrive. Perhaps the silver birch seed needs different amounts of light, water and nutrients to germinate and to develop into a sapling than the coco-de-mer palm nut. This might get you thinking: why do different species of tree have such different-sized and -shaped seeds, and do seed size and shape enable some species to thrive in some environments but not in others? Scientists have addressed these questions, and by applying the scientific method they have been able to explain why one species has particularly giant seeds.

The coco-de-mer palm is a species of nut-producing tree found on only two islands in the Seychelles. It produces the largest seeds of any plant. Standing beneath a coco-de-mer palm when its nut is ready to fall is foolish. Each can weigh up to 20 kilograms and measure a third of a metre across; a direct hit on the head could kill you. By way of contrast, you would likely not notice if the seed of a silver birch tree hit you on the head. Each seed is 1–2 mm across and weighs only a fraction of a gram. Unlike the coco de mere, the silver birch is a tree that has conquered more than two islands – it is found throughout much of Europe and Asia. Biologists have not paid the diminutive size of silver birch seeds much attention, but they have been intrigued by the large size of the seeds of the coco-de-mer palm for years, and we now know why they grow so large.

The earliest palms evolved 80 million years ago, about 14 million years before the extinction of the dinosaurs. Since the evolution of the first palms, the coco de mere species has evolved to thrive in habitats with poor-quality soil where minerals and nutrients are scarce. The species has evolved characteristics that make it a specialist at doing well in harsh environments where individuals of other species are usually doomed to an early death. The cost of being able to live in such poor environments is that the coco de mere is outcompeted by other plants in higher-nutrient soils. What characteristics has the coco de mere evolved to be a specialist in low-nutrient conditions?

Evolution has shaped the leaves of the coco de mere to

act like gutters, channelling water to fall near the trunk of the tree. When rain falls, it washes off bird poo, pollen and other bits of detritus that the leaves have accumulated. The rain, now laced with valuable nutrients and minerals deposited on the leaves, is channelled to reach the soil near the tree's trunk. The minerals and nutrients soak into the soil, to be absorbed by the tree's roots and used to grow new branches, leaves, fruits or more roots. The plant has evolved uniquely shaped leaves to capture nutrients and minerals that are scarce in the soil in the environment where it lives.

The coco de mere is also remarkably efficient at extracting nutrients from its old leaves before they are shed. No other plant species has been studied that is so good at reusing nutrients. The environment where the coco de mere lives is so harsh that all resources must be recycled if they possibly can be. Nothing should be wasted. But that does not explain why the seeds are so large.

The palm invests these valuable resources into huge nuts that take up to four years to develop, and two years to germinate once they have fallen. The coco de mere has evolved large nuts that are packed with nutrients to give germinating seedlings a great start in life. Seedlings of the coco de mere are so well provisioned by their seeds that they can quickly grow to a size where they can capture rare minerals and nutrients via their own gutter-like leaves. They start life with a built-in advantage over species with smaller seeds that cannot grow so large with the nutrients their parent tree has provisioned them with.

Large seeds do not fall far from the parent tree, and that is the case of the coco de mere, with most seeds germinating in the shade that their parent casts. What is truly remarkable is that coco de mere are the only plants known to nurture their young. Many animal species do this, but only one plant species appears to. Experiments have revealed that seeds which fall underneath their parent tree grow faster than those that grow further away. Part of the reason for this is that the guttering effects of the parent tree's leaves channelling water and detritus close to its trunk also provide benefits to its nearby offspring. But that is not the whole story. Many trees actively produce chemicals to hinder seeds of their own species germinating under their canopy to reduce competition. Coco de mere do not do this, but rather parent trees provide an environment that actively helps their young.

Instead of only producing one or a few seeds each year, like the coco de mere, silver birch trees can produce hundreds of thousands of seeds annually. These tiny seeds have little wings, and they can be carried by the wind for many kilometres. Silver birch seeds rarely fare well under the forest canopy; instead they thrive when they end up in an open environment with little competition. This means that very few of them arrive in suitable conditions for germination. The lifestyle of the silver birch is consequently very different from that of the coco de mere. One produces millions of seeds, of which nearly all are expected to meet an early death, while the other produces very few seeds over the lifespan, but each seed has a reasonable chance of making it to become an adult. Plant biologists have used the scientific method – observation, hypothesis, experiment and repeat – to gain an impressive understanding of why particular species are the way they are, and why they have particular adaptations to thrive in the environment in which they are found.

You might never have thought about why a particular tree is where it is, but if someone had asked you to explain it, you would likely come up with hypotheses as to how its seed got there and why it germinated and developed into a sapling and then an adult. You could do this because at heart we are all scientists. We all construct hypotheses to explain aspects of the world around us. These hypotheses might include something as trivial as why someone blanked you in the street this morning, through to something as monumental as what is the meaning of life. Often you will have competing hypotheses to explain a particular observation. You don't know why the dog vomited on the carpet, but it might be because he ate a discarded kebab in the park on his walk yesterday, or because he had a rabies jab two days ago.

Posing hypotheses is the second step in the scientific method after having made observations. The next step is collecting information to test the hypothesis. How is this done? What is good evidence, and what is not? How much evidence is enough? The scientific method involves more than observations, hypotheses and experiments. It involves knowing that your observations or experimental results are real and not just due to a chance result. How do scientists weigh the evidence and decide whether a hypothesis is supported or not?

Bold claims need strong support – if you claim to have a cure for cancer you need to be able to demonstrate it in a way that convinces other scientists. Part of the scientific method is being able to provide evidence to support your conclusions, and for other people to be able to repeat your work and get the same results. It is all very well to be able to come up with a hypothesis, report some observations or design an experiment to test it, only to discover that the data you collected are insufficient to test your hypothesis in a convincing manner. Let's consider three examples of human-like species as examples of different levels of evidence.

Leprechauns are supernatural solitary fairies from Irish folklore. They are short in stature and – the males at least – are often depicted as being bearded and dressed in green. I might tell you, even in a way that sounds convincing, that a leprechaun inhabits my garden, but until I can provide evidence of his existence, you should not believe me. Just saying stuff isn't good evidence. What evidence would you want? Ideally you would like to meet the leprechaun yourself or see his body, if he were to pass away. You might accept high-quality, undoctored photos or a genetic sample of a new species related to humans, but you should not take my word for it, however convincing I might sound. The great strength of science is it is evidence-based, and evidence needs to be produced in support or refutation of hypotheses. If hypotheses are supported in very large numbers of experiments, these hypotheses become facts. For the record, I have no evidence of a leprechaun living in my back garden. But that is no surprise, as how could he survive being trampled by my herd of unicorns?

You would probably think any leprechaun claims I made were far-fetched, but have you ever heard of the *Orang pendek*? Every now and again someone reports having found evidence of a new, extant, human-like species. Someone returns from an expedition with 'evidence' of the yeti, Bigfoot or the *Orang pendek*: a bipedal ape that is alleged to live in remote mountainous forests in Sumatra. The same pattern of events is then repeated. The returning adventurer presents their evidence of the new species, be it a sasquatch, skunk ape or batutut, and a media frenzy follows. The story quickly blows itself out and the media move on to a misbehaving politician or celebrity – anything else that can fill the 'and finally' slot of news programmes. Once out of the public spotlight, it gradually becomes apparent that the evidence in support of the new species is weak: a grainy photograph that could be of pretty much anything, a dubious plaster cast of a footprint, a hair that is thought to belong to the beast, or some unverifiable accounts from members of the local population. In some cases, the hair may be examined by geneticists who conclude it comes from a cow, goat or dog, or they are unable to extract any DNA (deoxyribonucleic acid, the molecule in which the instructions to build an organism are encoded) at all from it. The person who collected the weak evidence is often convinced of what they saw, but they do not always understand that most scientists require more compelling evidence, and ideally a specimen such as a skeleton from an individual of the species.

I once attended a talk describing insights into the *Orang pendek* at the Institute of Zoology, the research arm of London Zoo. The speaker was convinced by the evidence they presented, but the audience was not. In discussion afterwards, it became clear that most people in the audience wanted the *Orang pendek* to exist. The discovery of a new great ape would be a major biological advance, and the discoverer would be rightly feted. However, most of the audience felt that the photos projected were too blurred to count as evidence, and they gave little weight to an interview with a member of the local community whose testimony was presented as evidence of the *Orang pendek*'s existence. The interviewee was clearly an expert in the native flora and fauna, having spent many years in the forest. But he was unable to return to the site where he claimed to have buried

34

an *Orang pendek* he had found dead, and his statement that the species could remove their feet and put them on backwards to confuse trackers lacked credibility, as no other apes or, for that matter, any animal species, can do this. It would be a strange thing to evolve, as such a trait is highly unlikely to give the bearer any advantage in producing babies, the currency of evolution, compared to a firmer-ankled individual. One member of the audience felt the interviewee's claim to have been a pigeon in a former life was more believable than the foot-switching trick. We wished the researcher luck in collecting more data while explaining that we were unconvinced and describing what would be required to persuade most of us. I still hope to hear one day of a specimen of a new species of great ape found to be living in Sumatra, but I am not holding my breath. I would also love to learn that sasquatch and yetis roam remote parts of our planet but, once again, I suspect it is unlikely.

You don't always need much material to provide strong evidence of the existence of a species. Leprechauns don't exist, *Orang pendek* probably don't exist, but woman X did exist, and she was from a group of individuals called Denisovans. The Denisovans were an ancient people, like Neanderthals, but they lived in Central Asia rather than Europe, and were closely related to us. Scientists have not found enough material to classify the Denisovans as a new species, but what they have discovered is remarkable. Much of the story we know comes from a single finger bone that was subjected to high-tech analyses which have shed a lot of light on the history of our own species.

The story starts in a cave in the Altai Mountains in Siberia, which used to be the home of a Russian hermit. The hermit, called Dionisij, made the cave his home in the eighteenth century. Denis is the English translation of Dionisij, and this led to the cave being named the Denisova Cave.

Denis was not the first human to use this cave. Russian archaeologists who have worked in the cave have found evidence that Neanderthals, 'woman X' and, more recently, modern humans, used the cave over a period of at least 100,000 years. Woman X was identified by a bone from the finger of a child that was neither Neanderthal nor *Homo sapiens* – the Latin name of modern-day humans. It was not the structure of the bone that revealed this but the DNA sequences that were extracted from the finger.

The girl whose bone it was used the cave sometime between 30,000 and 50,000 years ago. Biologists are yet to conclude whether the girl is a different species to us, or whether she is from the population of a subspecies, and this means that her scientific name is yet to be decided. Biologists classify organisms as belonging to species, but what defines a species can be challenging because individuals of the same species can vary quite substantially among different populations across the species range. For example, the Bengal tiger (*Panthera tigris tigris*), the Sumatran tiger (*Panthera tigris sumatrae*) and the Siberian tiger (*Panthera tigris altaica*) all look a little different from one another, they are genetically distinct, and their populations have evolved away from one another for potentially hundreds of generations. Nonetheless, they are all tigers and can mate with one another to produce viable offspring. They are consequently described as subspecies. Woman X is usually referred to as a Denisovan, and if I were asked to make a call, I would classify Denisovans, Neanderthals and modern-day humans as subspecies of one another rather than as separate species, but this may irritate some palaeontologists.

We know little about woman X's life, but DNA has been extracted from her finger bone and sequenced, with great care, using state-of-the-art technology. The genome from this girl has been compared with genomes from modern-day humans from around the world, and with Neanderthals. DNA revealed she was not a Neanderthal, nor a modern human. Woman X's passing may have been unremarkable, although I have no doubt she was mourned by her relatives and friends, but the discovery of her finger bone has left an astonishing legacy: scientists have had to revisit the history of us.

The discovery of Denisovans caused much scientific excitement, but it was not the first time evidence of an extinct subspecies so similar to us caused interest. In 1856, three years before Darwin published *On the Origin of Species*, bones were discovered in a cave in the Neander Valley in Germany. They clearly differed from those of modern humans, and they caused excitement when the find was announced in the Elberfeld newspaper. Over the following years learned researchers argued that the bones were from a Russian Cossack, a Native American, a soldier from Attila's army, a man who suffered from and was pained by rickets, or a member of an extinct tribe of primitive villagers that inhabited Europe before modern people arrived. William King published a paper in 1864 that drew similarities between the fossilized bones and those of apes like chimpanzees and gorillas, and proposed that they belonged to a new species he called *Homo neanderthalensis*. Scientists began to accept that another species (or subspecies) with similarities to humans once existed in Europe, and the Neanderthals were generally assumed to be simple, brutal savages. It took most of the following 150 years for scientists to overturn this view.

Until the ability to sequence tiny amounts of DNA from ancient bones, the standard view of human evolution was that a species called *Homo erectus* evolved in Africa before dispersing into Europe and Asia 2 million years ago. The species was astonishingly successful, evolving into Neanderthals and various other species including the diminutive *Homo floresiensis*, or the hobbit-man of the Indonesian island of Flores. In Africa, descendants of *Homo erectus* evolved into modern-day humans about 300,000 years ago. The first *Homo sapiens* dispersed out of Africa 125,000 years later, but they only got as far as the Arabian Peninsula, Syria and Turkey. They failed to establish permanent populations out of Africa and subsequently died out. But 60,000 years ago a second out-of-Africa migration occurred. These *Homo sapiens* were more successful, reaching South Asia between 50,000 and 60,000 years ago before making it to Australia a few thousand years later. Europe was successfully colonized around 40,000 years ago. America was colonized after Europe, sometime between 25,000 and 16,500 years ago. As *Homo sapiens* spread across the globe, they outcompeted the more primitive hominin species that evolved from *Homo erectus*, with Neanderthals being the last species to go extinct, about 30,000 years ago.

By the mid-1990s, speculation on whether humans dispersing from Africa had bred with Neanderthals had become rife, but in the absence of fossils of copulating pairs of Neanderthals and humans the debate was evidence-free. I remember that at some point in 1994 a colleague of mine was asked during a TV interview whether our ancestors mated with Neanderthals. He explained that there was no evidence but thought probably not. A few days later he received a letter from a viewer who wrote, 'Neanderthals and modern humans have reproduced, and I have evidence. Please find enclosed a photo of my partner.' An amusing anecdote, but given that Neanderthals died out many millennia ago, it seemed back in the 1990s that we might never know whether they interbred with humans. It is remarkable how genetics has advanced the field of human evolution in the last three decades.

The discovery of woman X's finger bone, along with finds of well-preserved Neanderthal remains, coupled with significant technological advances in extracting DNA from ancient fossils and assembling genome sequences from the fragments, allowed biologists to address the question of interbreeding between Neanderthals, *Homo sapiens* and Denisovans in a scientifically rigorous way. But the work had to be conducted extraordinarily carefully.

Over time, DNA in dead bodies degrades as the chemical bonds between the atoms from which it is made break. The rate at which it breaks down depends upon the environment. The average temperature in the Denisova cave since the death of woman X has been about freezing, and the relatively dry air in the cave meant some DNA from woman X had survived and scientists were able to extract tiny amounts of DNA from the finger bone. Having done this, they had to be careful that the phalanx (the scientific name for a bone from a finger or toe) had not been contaminated with DNA from other people, including Dionisij and the scientists who discovered it. They also had to be certain the DNA they extracted belonged to something closely related to humans, and not to any other animals or bacteria that may have used or lived in the cave. By following extremely careful procedures to ensure the DNA was not contaminated with DNA from anywhere else, a group of biologists based in Germany was able to read, or sequence, the DNA of woman X. They could then compare her DNA with that of lots of modern humans living in different parts of the globe, and with DNA obtained from Neanderthal bones and teeth.

Although the amount of material used in the initial analysis of woman X came from a single finger bone, it generated a lot of data. DNA consists of strands of molecules called nucleobases, which come in four varieties called adenine, guanine, cytosine and thymine, or A, G, C and T. The order of bases in a strand of DNA is known as a DNA sequence, with AACACTGT being a different sequence from ATTAGAGC. Your genome consists of about 3 billion nucleobases linked together in 46 different strands, with a strand referred to as a chromosome. The genome of woman X was initially sequenced to a depth of 1.9 times, which means that, on average, each nucleobase on each chromosome was recorded just under two times. Any single A, G, C or T within a strand was recorded by a machine designed to read strings of nucleobases. Each nucleobase may have never been read, or may have been read a few times, but the average number of times each nucleobase was read was 1.9. But why read each nucleobase more than once?

When genomes are scored, it is not currently possible to read the order of nucleobases across an entire chromosome in one go. The reason for this is that sequencing involves taking a specimen's DNA from multiple cells, and results in each chromosome being chopped into fragments. Many of these fragments are short and might only be about a few tens of nucleobases long. Genetic sequencing involves reading the order of nucleobases in very many fragments of DNA before using computer algorithms to stitch them together.

The larger the number of times each nucleobase is scored along with its neighbours on different fragments of DNA, the greater the confidence that the gene sequence of a chromosome can be correctly assembled from all the different fragments.

You might wonder why this is so complicated. Imagine a very short sequence consisting of ACAGTCAGA which is split into two fragments, ACAG and TCAGA. How should these be joined? ACAG first and then TCAGA, or TCAGA and then ACAG? Things are even more complicated because we don't know whether each fragment represents DNA going from left to right or from right to left in the genetic code. Should it be ACAG or GACA? With a depth of 1.9 times, it is hard to stitch together a whole genome with great confidence. These days it is not unusual to read of genomes being sequenced to a depth of 30 times, and indeed subsequent genomic work on woman X five years after the first study was published achieved close to this.

A useful analogy to help understand genome sequencing is to imagine taking multiple copies of this book, shredding each one, such that you have many fragments of a few words each, then trying to reassemble the book. Except it is much harder with genetic sequences than with a book because you only have four letters, rather than the twenty-six of the English alphabet, no spaces, and, unlike with words, you do not know whether to read each fragment from left to right, or vice versa. Furthermore, there are just over three quarters of a million characters in this book, while genomes of humans and our ancestors are much longer. Constructing genetic sequences is hard work.

The genome size of Neanderthals, humans and Denisovans is each about 3 billion base pairs. The scientists who genotyped woman X compared genome sequences between the Denisovan, a Neanderthal, twelve humans from different parts of the world and a chimpanzee. That is a huge amount of data: about 45 billion nucleobases. These comparisons revealed sections of Denisovan DNA in modern-day humans living in parts of South Asia and Oceania, while Neanderthal DNA snippets were found in modern-day Eurasian populations.

What are you most likely to trust when considering whether species similar to humans exist or once existed? My word for the existence of leprechauns, a grainy out-of-focus photo and the testimony of a man who used to be a pigeon for the *Orang pendek*, or 45 billion nucleobases and some extremely careful genome sequencing and formal statistical analyses for woman X? Science is evidence-based, and the evidence that our ancestors mated with Neanderthals and Denisovans is extremely compelling. Despite this, there is still the odd scientist who is attempting to rule out other processes, such as biased patterns of genetic replication errors (mutations in DNA), that could potentially generate the same patterns in data, but few people find these explanations compelling. Nevertheless, posing alternative hypotheses is important. Science is all about posing and testing hypotheses to identify those that can produce an observed pattern.

It is not just the history of humans that has benefited from technological advances that can create vast amounts of data. The particle collisions which happen at the LHC generate so much data on the fundamental particles that make up the universe that only a fraction of the data – the fraction that looks like it might reveal exciting new science – is stored, and the rest is discarded. Space telescopes like the recently launched James Webb Space Telescope also produce vast amounts of data. The orbiting telescope sends back to Earth large numbers of high-quality images of galaxies billions of light years and trillions of miles away. You, too, create lots of data, as your phone communicates with phone masts, and as it records your steps or the number of stairs you climb. Your social media posts, and the posts you look at, generate networks of views, and the transactions you make with your credit and debit cards are all creating data. It is estimated that humanity generates 1.145 trillion megabytes of data every day. To give you an idea of quite how much information that is, I can save the 130,000 or so words that constitute this book in just under half a megabyte.

The trend for ever more data will likely continue, with humanity doubling our amount of data every eighteen months to two years. Back in 1945, the doubling time used to be twenty-five years. No one can keep track of all these new data being produced – data is plural, datum is singular – and analysing even bits of it is a challenge that is being met via the emergence of a new field of science: data science. If you can wrangle and analyse data you will be in demand in jobs as diverse as banking, political influencing and gaming. As the range of equipment available to scientists to make ever more detailed observations increases, so too will the rate of data generation, and quite possibly human knowledge too.

To collect data, you need something to observe and measure, whether that material is a DNA sequence, the debris from particles colliding at the LHC or a 125-million-year-old *T. rex* fossil. There are more modern-day things to measure than there are objects from the past. We know more about the life of Thales of Miletus than we do about woman X, in part because woman X lived thousands of years before Thales was born. She was born before the first cities, and before writing was invented and records were kept. But detailed analyses of information written in her DNA have provided us with a picture of part of the history of humans. Thanks to our ability to extract genetic samples from woman X's finger and to our understanding of the general rules of how genetics work, we have been able to piece together a picture of how humans evolved. Although the hypotheses that are tested are much more complex than those concerning the presence of a particular tree in a specific location, they are all based on the scientific method. This is why the scientific method is humanity's greatest achievement. Not only can it answer questions about our history, about the location and adaptations of trees, but also about the history of the universe.

Compelling observation is key to good science, and we have got very good at making detailed and diverse observations. But it is not the whole story. All these data create challenges. Not only can analysing data require a lot of computational resources, but how strong does a pattern have to be for us to be confident it hasn't just arisen by chance, or that the pattern is something we should worry about?

Of my undergraduate students, 61.4 per cent do not like statistics. Convincing most people that statistics is fun, or useful, can be far from straightforward. But statistical analysis is one of the cornerstones of modern science. The aim of statistics is to provide some measure of confidence that a pattern in data is real.

Imagine you have a large bag of sweets. You dip your hand in and remove one sweet. It is red. Does that mean that all the sweets in the bag are red? Intuitively you know you cannot conclude that because you have only looked at – or sampled, as statisticians would say – one sweet out of perhaps several hundred. If you sampled two sweets from the bag and they were both red, you might start to be a little more confident that you have a bag containing only red sweets. Your confidence would grow that you had a bag of red sweets if you sampled fifty sweets and they were all red, but you would still not be 100 per cent confident that all sweets in the bag were red. You can only be 100 per cent confident if you have examined all the sweets and you found that they were all red. What statistics does is provide a way to put a measure of confidence in any patterns you might find in data. In this case, your data is the number of red sweets you have pulled from the bag, and the pattern is they are all red. You want to be able to say something like 'I am 85 per cent confident all the sweets are red.' In the same way we are all scientists at heart, we are also all statisticians. We are endlessly making decisions based on our confidence that something will, or will not, happen. As I wrote this, I was gambling someone would want to publish this book.

It was a biologist, Ronald Fisher, who invented modern statistics. He was born in England in 1890 and died in Australia in 1962, aged seventy-two. The list of statistical tools Fisher developed is remarkable, and anyone who has ever had to sit through lessons on statistics can thank Fisher because he invented things with names like 'maximum likelihood', 'analysis of variance' and the 'F-test'. These are still taught in undergraduate science, medicine and social science courses.

Since Fisher laid the foundations of statistics, the field has advanced considerably. There is now a plethora of exotic-sounding statistical methods including reversible-jump Markov Chain Monte Carlo, trans-dimensional simulating annealing, and hierarchical Bayesian multistate modelling. All these methods stem from Fisher's fundamental insight that variation in the observations scientists make can be broken down into contributions from different sources. To illustrate this logic, imagine you measured the weight of everyone who lives in your neighbourhood. Different people will have different weights, which means there is variation in weight within the population you have collected the data from. Statistical methods allow you to explain where this variation comes from. Is it due to diet, where you were born, your genes, how much exercise you do, your gender, your height or your age? Fisher developed methods to estimate how factors such as these contributed to variation in data, and to assign a measure of confidence in the numbers produced.

Depending on how much data has been collected, the degree of confidence that is considered acceptable can vary. In physics, the level of confidence frequently used is called five sigma, and what this means is that there is about a one in a million chance that the result happened by chance alone. In other fields, where fewer data exist, such as palaeontology, the level of confidence that is deemed appropriate to discuss a finding as likely being real and not due to luck is one in twenty.

Statistics is a way of putting confidence in any patterns found in data, and statistical tests can be applied to both observational and experimental data. When a pattern is identified in observational data, the next step, if possible, is to construct an experiment to explore whether a process or mechanism hypothesized to generate the pattern is plausible. Revisiting my 'why is that tree there?' discussion from earlier in the chapter, I might plant seeds from a range of trees in a range of soil conditions and see what germinates. Good experiments have various treatments and a control. Different seed types, and different soil types, would be the treatments in this example, while the controls might be no seeds planted, or seeds without soil, or indeed no seeds or soil to explore whether the spontaneous emergence of life forms can occur. (It can't.)

A good experiment generates new data to test a hypothesis on the cause of a pattern seen in existing data. These new data can then be subjected to more statistical analysis. If the experiment provides support for a hypothesis, we end up at a point where we have seen some pattern in data, posed some hypotheses as to what generated that pattern, and generated data from experiments that test which of the hypotheses is plausible. If a hypothesis is not supported, this is useful information, and could lead to us devising a new experiment or set of experiments.

In some cases, experiments are just not possible. For example, we are unable to re-create conditions seen at the birth of the universe as it is beyond our technical expertise. A consequence of this is that we must rely on observations made by telescopes of distant objects in the night sky. When experiments are impossible, the next best thing is to build a mathematical model and see whether it generates predictions that match observations. Scientists have built models of the development of the universe, and they examine the outputs of each model to see which look most similar to what we observe. The models that most closely match reality are most likely to capture processes that generated the patterns we see.

Mathematics is an exact and unforgiving language. It is abstract and describes how different things are related to one another, and how they change over time, or across space, or even in abstract dimensions that we cannot really envisage.

A bit like many human endeavours such as sports, music, art or learning to speak other languages, there is a huge amount of variation in mathematical ability between people. To some people it is all obvious, while for others it will always be gobble-degook, however much effort they invest in trying to master it. Not all science requires the use of mathematics. Many superb experimentalists, fieldworkers and lab-based researchers are brilliant scientists yet poor or mediocre mathematicians. I am reasonable, but not exceptional at mathematics, but I am hopeless at fieldwork. I do tend to work with collaborators who are experts in fields where I lack skills, and one of my skills is being able to bring people with complementary skill sets together to address interesting questions.

There are mathematical models not only of the development of the universe but also of the behaviour of solar systems and galaxies, of chemical reactions, of the diversification of life and of the functioning of the human brain. I cannot think of an aspect of science that has not been modelled using equations. Much like there are different genres of music and art, there are different genres of mathematical models. Some models are kept deliberately simple to examine how one process might create a particular pattern, while others are highly complicated, providing real-time, accurate descriptions of the system under study. Each genre of model is useful in some way, and the trick is knowing what sort of model to use. Albert Einstein, one of the greatest scientists ever, stated that 'everything should be made as simple as possible, but not simpler', and this has sometimes been interpreted to mean that models should be kept super-simple, but that is not an appropriate interpretation. Sometimes models need to be complex to achieve their aims. For example, imagine you developed a complicated mathematical model that perfectly predicted the stock market. You would be extremely happy. If each attempt to simplify the model resulted in its predictive accuracy disappearing, the model no longer does what you want it to do, and you conclude the full complexity of your model is necessary to allow early retirement. You probably wouldn't care if you didn't understand why your model worked, you'd just be happy that it did. In this case, your model's complexity is at the right level because it does what it was designed to do: make you rich.

Models are also an extremely important tool in generating new hypotheses. Once a model is constructed, it can be used to make predictions, and these predictions can sometimes be tested with new observations or experiments. The Standard Model of particle physics is a complicated equation that describes the workings of three of the four fundamental forces of the universe: the weak and strong nuclear forces and electromagnetism. Analysis of the model led to prediction of a particle called the Higgs boson. The LHC, the world's largest machine, was constructed to search for the boson, and it found it. A mathematical model predicted a new fundamental particle and, when scientists looked, it was where the model said it would be. That is an impressive use of a mathematical model to pose a hypothesis that experiment proved was supported.

The LHC is spectacular. It is an example of the remarkable technological progress that scientists and engineers have made over the last couple of centuries. We can now make new types of measurements that were unthinkable even just a decade ago, and we can also now measure things much more precisely and with much less error than ever before. We have also learned to manipulate nature in new ways, altering some aspect of it and seeing what the outcome is.

Genetics is one area where our ability to understand, and alter, nature has increased dramatically. When I was at university in the late 1980s, the genetics we did almost entirely revolved around fruit flies, *Drosophila melanogaster*. We would take flies with different characteristics, breed them and then score the traits in offspring. The characteristics would be things like eye colour, or the number of bristles on their legs. It was dull work, but we were able to infer inheritance patterns of the traits and work out how many genes were involved. Although I always enjoyed statistics and evolution, I found these practical sessions uninspiring. But I did get taught something, and I now work on a project where we have been able to take things much further than studying flies that look a little different from one another. It is also a project where a mathematical model provided new understanding.

One of my study systems is the population of grey wolves living in Yellowstone National Park. It is a wonderful place to work, and the reintroduction and re-establishment of wolves in and around the park is one of the great conservation success stories, thanks largely to the tireless work of my friend and collaborator Doug Smith. Not everyone likes wolves in their backyards, and despite the improvements to the landscape due to wolves suppressing elk numbers, there are various groups that do not see it this way. For example, hunters now have to get out of their cars to shoot an elk and for that reason they dislike wolves.

Although the common name for the species *Canis lupis* is the grey wolf, wolves can have either a black or grey coat colour. The genetic variant that causes the black coat colour evolved in domestic dogs and was passed on to wolves when they mated with dogs shortly after people arrived in North

America. Black wolf numbers then started to increase. Black wolves are not found everywhere in North America but tend to increase in number from the far north-east near Nova Scotia, where they do not occur, to the south-west edge of their range towards Mexico, where they are common. Elsewhere in the world, black wolves are absent or extremely unusual.

If we had conducted the fly experiments I did at school and university with wolves, we would have learned that coat colour is caused by genetic variants at one gene. If a grey wolf mates with another grey wolf the pups that are produced are always grey. If a black wolf mates with a grey wolf, or if a black wolf mates with another black wolf, the young can be either black or grey. The frequency of the black and grey pups reveals that there are two variants, or alleles, at the gene; let's call them Grey and Black, or G and B. Each wolf has two alleles. If they are both G, the wolf has a grey coat, and if they are both B, the wolf is black. If one is G and the other B, then the wolf is also black. The B allele is said to be dominant over the G one. I would have really enjoyed doing a wolf breeding experiment at school rather than working with fruit flies, but health and safety regulations forbade it.

Across their range, black wolves tend to be more common in areas where there are many forests, so a hypothesis was proposed that black wolves had an advantage over grey ones when hunting in forests, as they were better camouflaged within the trees. We conducted statistical analyses of survival and reproductive rates among wolves with different genes and found something intriguing. Black wolves with one B allele and one G allele – BG wolves – were better at surviving and reproducing than BB wolves. These black BG wolves were also slightly better at surviving than GG wolves. What this finding meant is that the black coat colour could not be giving the wolves an advantage while hunting in forests, but genetic variants at the locus that made them black or grey must be responsible for something else. The reason we can conclude this is that although BB wolves and BG wolves look identical they have very different lifetime prospects at birth. If being black were advantageous, then the BB wolves should survive as well as the BG wolves. Yet BB wolves tended to die young, and if they did survive, they rarely reproduced. In contrast, BG wolves thrived. If being black were an advantage, then both types of black wolf should be equally good at surviving and reproducing. Because they are not, the alleles that determine a wolf's colour must do more than just determine their colour.

The gene that makes wolves black or grey is called CBD103, and we know that in other species this gene is associated with immunity to diseases. So we posed a new hypothesis: that BG wolves were better at avoiding or fighting off infections compared to BB or GG wolves. Every few years in Yellowstone a disease called canine distemper virus flares up. Canine distemper virus, or CDV, is a bit like measles for wolves and other carnivores, and we hypothesized that black BG wolves were less likely to die in years when CDV outbreaks occurred compared to black BB or grey GG wolves. Statistical analyses supported this hypothesis. The difference in survival rates between BG and GG was much less pronounced in years when CDV did not break out, while BB wolves always fared badly.

CDV can infect many species of mammal including bears, racoons and skunks. The disease only persists in an area if there are many species of mammal that can catch the disease. There tend to be more of these species towards the south-west of the wolf's range and fewer to the north-east. The presence of black wolves was consequently determined by the presence of CDV, and the presence of CDV was determined by the number of species that could become infected by the disease in an area.

Although it was very exciting that our hypothesis was supported, there may be other explanations for the patterns we observed. Because of this, some colleagues of mine in Los Angeles led by Bob Wayne, who sadly recently died, designed a remarkable experiment which shows how technology has revolutionized the study of wild animals.

To study the wolves, we need to find them, and to do that we attach collars that transmit radio signals to one or two wolves in each pack. The collars do not hurt the wolves and are like a chunky collar that a dog might wear. To fit a collar to a wolf involves darting them with a tranquillizer dart, and while they are asleep we check them for diseases, weigh them and take a blood sample and a scrape of cells from inside their cheek. These cells are frozen before being flown to Los Angeles, where my colleagues would grow them in the lab. Each wolf that we have caught consequently had a cheek cell culture in the lab in Los Angeles. Using some clever genetic engineering methods called CRISPR, my colleagues created three cheek cell lines for each wolf: one with the BB genotype at the gene CBD103, another one with the BG genotype, and a third one with the GG genotype. No other genes were altered. A few of these genetically engineered cells were then exposed to pathogens and the way they responded was recorded. The experiments were designed to test whether it is the BG genotype at the CBD103 gene that provides an advantage, or some other genes. The results revealed that the genotype at CBD103 does play an important role in immunity, along with many other genes.

Although the gene editing technology is remarkable, and statistical analyses revealed different survival rates between wolves with different genotypes in years with and without CDV outbreaks, we did not know whether these survival differences had any real impact on the wolf population. Everything has to die, so do different rates of death by CDV in wolves of different genotypes matter in any way? To address this question, we constructed and analysed a mathematical model. Analysis of our model led us to an interesting prediction. Assume, for a moment, that you are a black wolf living in Yellowstone where CDV breaks out every few years. The best strategy for you to follow, from an evolutionary perspective, is the one that maximizes the number of copies of your genes in future generations. In the case of the wolves, the best strategy would be to produce black BG offspring (the BG genotype, remember, confers immunity to CDV). What is the best way to do this? It turns out that if you are a black wolf, you should mate with a grey wolf, and vice versa. Regardless of whether you are a black BB or a black BG wolf, the way to maximize your chances of producing BG pups is to choose a mate of the opposite colour, and the same is true for grey GG wolves. Wolf pairs with different coat colours should occur more frequently than wolf pairs with the same coat colour. We tested this prediction by looking at who mated with whom in Yellowstone and we discovered that black–grey matings occurred more often than would be expected if wolves mated randomly. On average, black and grey wolves are more likely to pair up than two black or two grey wolves. However, our model also predicted something else. In the absence of CDV, we predicted it would always be better to produce grey GG wolves. This is because grey wolves are evolutionarily fitter than black ones when in CDV-free locations. Grey wolves should mate with grey wolves, and the Black allele, and hence black wolves, should eventually disappear from the population. This is exactly what we found. In areas of North America where CDV outbreaks occurred only vary rarely or not at all, there are no black wolves.

The predictions of our model are not as impressive as the Standard Model of particle physics that predicted a new particle that was subsequently found. But our model made predictions we could test, and those predictions were supported. Our model helps show what a remarkable force evolution is. In areas where CDV occurs, those wolves who, by genetic chance, had a predisposition to favour wolves of a different colour were more likely to produce surviving pups than those who chose mates that were the same colour as themselves. Over time, in areas where CDV outbreaks occurred, an ever greater number of wolves mated with individuals of the opposite colour, and this has produced the distribution of black wolf occurrence seen across North America.

Technological advances have not only revolutionized genetics but also physics, chemistry and computer science. We can now observe galaxies over 13 billion light years away, as well as the structure of molecules a few atoms across. We can measure very faint gravitational waves formed by distant black holes colliding, and we can accelerate some of the fundamental particles of matter to 99.999999 per cent of the speed of light using electromagnets. Artificial intelligence can now fool us into thinking we're communicating with another person rather than a computer, and write passable undergraduate essays. Our technological journey began when our ancestors first knapped stones to create sharp edges. We have come a long way since then, and our latest technology is both astonishing and a little concerning.

The ability to edit genotypes at a gene in individual cells as we have done in the wolves offers potential for treating debilitating diseases in humans if society decides it is happy for such tools to be used in our armoury of approaches to treat illness. My colleagues' research using wolf cheek cells helps show what is possible, and the results reveal what can be achieved. But we must treat such technology with respect. We currently do not have sufficient understanding of the human genome to genetically engineer individuals with resistance to disease, or particular skills or abilities, but knowledge is growing rapidly, so it is important that the use of such technology, including gene therapy, is appropriately regulated. Similar concerns have been raised about the latest generation of artificial intelligence technology. ChatGPT is undoubtedly a useful tool, but we must make sure we use it only for good.

Observations describe a pattern, processes chronicle why the pattern happens and mechanisms explain how processes occur. For example, the observation that hydrogen and oxygen atoms combine to form water molecules is a pattern; why the atoms behave like this is because the oxygen and hydrogen atoms have opposing electric charges that attract them to one another to form a molecule; and how they bond together is because the atoms share particles called electrons. A detailed understanding of why a pattern is observed requires knowledge of both process and mechanism.

Some mechanisms are extremely complex, especially those involved in complex life forms. The emergence of the two separate lineages that would eventually result in chimpanzees and humans from a shared common ancestor that lived between 6 and 8 million years ago is a pattern; the processes that caused the diverging lineages is evolution by natural selection; while the mechanisms that resulted in separate chimp and human lineages are errors, or mutations, in the genetic code that impacts the way individuals of each species develop. These contrasting developmental trajectories involve different genes being turned on and off at different times in the growing embryo, different proteins being produced in some of our cells, and these proteins impacting whether we grow up looking like a person or a chimp. In this case, the pattern is well characterized, the process is well understood, and bits of, but by no means all, the developmental mechanisms have been characterized. Descriptions of the details of complex mechanisms quickly become very complicated, and in this book I largely avoid them. Instead, I focus on the processes that led from the Big Bang to me and you, giving overviews of the mechanisms only when this does not require technical detail.

Sometimes as science progresses there can be multiple interpretations of patterns. When this happens, and while researchers strive to collect more data that might allow one of a set of competing hypotheses to be supported with others rejected, most scientists turn to Occam's razor to decide which hypothesis to favour. William of Ockham was an English theologian and philosopher born in the village of Ockham in Surrey in 1287. His razor states that when there are multiple explanations for a phenomenon, go with the simplest. That is the approach I take when equally plausible explanations have been proposed, and in particular I use this philosophy in the final chapter when I answer the question of whether our existence was inevitable or down to chance.

As well as answering that question, I want this book to convince you of science's remarkable achievements. I can only summarize a tiny fraction of scientific understanding, but it is sufficient to provide a history of why we exist. Many people are wary of science, but they should not be. The history and workings of our universe are worth understanding, for they are both remarkable and beautiful. The scientific method should not be feared, it should be embraced, for the knowledge it has uncovered through the hard work of millions of scientists is breathtaking. Let's begin exploring the history of why we exist.

The Beginning

I have always been an avid reader. My parents would have to tell me to stop reading and to go outside to play, so the opposite of any normal kid, and I would hide under my covers with a torch and a book after being told to go to bed. I can't remember the first science book I read, but I would go to the mobile library that visited our village and borrow books on any aspect of science I could find. The best reads were often about physics, and particularly those that revealed secrets of the universe, although books on dinosaurs came a close second. I remember hoping scientists might even discover dinosaurs living on other planets.

Although I loved reading about physics at home, it was taught in a fabulously dull way at school. The curriculum required us to learn about the differences between potential and kinetic energy for what seemed like weeks on end, yet we were never taught about exciting things like supernovae. We would work out the angle a ball would bounce off a wall, but we never learned about particles such as quarks or photons. At home I would read about how time slowed down as objects neared the speed of light, but in the classroom the most exciting thing we did with light was to build a pinhole camera.

My teachers strove hard to make the lessons exciting, but they fought a losing battle, for whoever had designed the curriculum was on a mission to make an interesting subject boring. Physics is the science that explains why the universe behaves like it does. It describes how fundamental forces work, how energy and matter (anything that can be weighed and takes up space) are related, and why particles interact with one another in the way they do. It now seems remarkable that my classmates and I weren't wowed with mind-bending knowledge. I was left nonplussed but undeterred. I continued to read about the exciting bits of physics at home, and soon came to realize that the inanimate world could be fascinating, and even kinetic and potential energy could be interesting. Nonetheless, for those readers who suffered the same physics curriculum as I did, that is the last you will hear of these types of energy in this book.

For you and me to exist, the universe had also to exist, but how did it come to be? We take it for granted that our universe has the right characteristics to harbour life, but was it inevitable that the universe would turn out this way at its birth? Could a universe have formed in which it was impossible for life to evolve? Scientists have found that if some of the characteristics of our universe were just a little bit different, then atoms, stars, planets, you and I could not exist. The first steps in our history will take us from a minuscule point of intense energy from which our universe grew, to fundamental particles, to the first atoms, and then to the first stars. There is still much to learn about the early days of our universe, but scientists have an astonishing understanding of the particles, energy and forces that determine how our universe behaves, and how these laid the foundations for life on Earth. Before we start this journey, I will introduce a few facts about our universe.

Today our universe is truly massive. No one knows quite how big, but scientists have estimated that it may well be over 7 trillion light years across. It would take a beam of light

7,000,000,000,000 years to travel that vast distance. One light year is just under 6 trillion miles – that's how far light can travel in a year – which makes the universe 42 trillion trillion miles across. From Earth, we can see only a fraction of this, and scientists call this the observable universe. It is a mere 93 billion light years across, but it is getting bigger. Later in the chapter I explain why, along with why we can't see all of the universe.

Distances that are measured in billions of miles don't make much sense given our daily experience on Earth. One of the longest non-stop flights you can take is from London, England, to Perth, Australia. It takes 17 hours, and the plane travels 9,000 miles. That is 0.0000000015 of a light year. Light can make the journey in less than $1/20^{th}$ of a second. The Boeing 787-9 Dreamliner that makes the London–Perth journey has a top speed of 690 miles per hour, which is very sluggish compared to light. It is also slow compared to our fastest machines. NASA's Parker Solar probe is the fastest object humans have produced. It achieved 0.05 per cent of the speed of light when it hit a speed of 330,000 miles per hour as it passed close to the sun, but that is still only $1/2,000^{th}$ of light speed. Light is very fast, and our universe truly vast.

To understand how our enormous universe behaves it is necessary to examine the behaviour of the tiny particles from which you, I and all matter are made, along with how these particles interact with one another. These particles, as best as we can measure them, have a diameter of about a thousand trillionth of a millimetre. Scientists refer to these particles as fundamental because they cannot break them down into smaller parts. These particles are the building blocks from which all things are made.

The behaviour of large objects like rocks, cars, aeroplanes, asteroids, the moon and other planets in our solar system is familiar to us. These are known as macroscopic objects because they are large compared to fundamental particles. Everything we experience in our daily lives, from other people, through books, to a glass of water, is macroscopic. Macroscopic objects move through time and space just like us. They do not exist as a blur, nor do they disappear from one location to mysteriously appear in another, as some fundamental particles, such as electrons in an atom, can do. In contrast, on the spatial scale at which fundamental particles interact, the microscopic scale, particles behave very differently. They are hard to pin down to a location, and this makes the behaviour of these particles appear deeply peculiar. For some of this chapter and the next, when I describe the behaviour of fundamental particles it helps to suspend disbelief a little. Reality as we know it is not reality at the scale of fundamental particles.

Up until the first half of the twentieth century, most scientists assumed the universe was constant in size, was unchanging, and had always existed and always would. But as telescopes increased in power, it became apparent that the universe was not constant but was expanding, and this suggested it must have had a beginning, and at that beginning our universe must have been very small. It has grown and changed a lot since its birth as the tiny dot of intense energy that it must have been.

Our universe started out tiny but is now vast. In every direction we look from Earth we see universe. We cannot see the universe's edge, but there is still a lot to look at. We can visualize what we can see by considering a hypothetical space voyage. Imagine being an astronaut on a spacecraft that is 400 kilometres from the surface of our planet – that's the orbit of the International Space Station. You look out of the window, and you can see our beautiful home world. Let us now start a journey away from Earth. As our planet recedes into the distance, the next object we encounter is the moon, which, like the space station, orbits the Earth but further away. We then start to encounter other planets in our solar system. We see a total of eight planets orbiting the sun – Mercury, Venus, Earth, Mars, Jupiter, Saturn, Uranus and Neptune – along with a dwarf planet called Pluto, and millions of smaller rocks. As we leave the solar system, the light from the sun dims the further we travel, until other stars shine brighter than our own. We will eventually start to pass some of these other stars, each with its own set of orbiting planets and rocks. In due course, the stars start to become less frequent, and we realize that our sun, now lost in a vast cloud of other stars, is just one of at least one 100 billion stars that are spinning around a central point. All these stars are in our galaxy, the Milky Way.

There are about a trillion galaxies in the part of the universe we can see. When we look at galaxies beyond the Milky Way through telescopes, we discover something interesting: most of them are moving away from us, and the further away a galaxy is, the faster it tends to be moving. If we were to reverse time, something that cosmologists have done in computer simulations of the universe, we discover that all the matter and energy that is now spread across the vast tracts of the cosmos must once have been compressed into a single point called a singularity. The singularity would have been smaller than the smallest of fundamental particles and would have been extremely dense and hot, and from this point emerged the universe in which sentient life evolved at least once, here on Earth. Physicists have shown that key to the way our universe behaves, including why it is expanding, are four fundamental forces that are central to our story. These forces determine how particles, atoms and larger objects including planets, stars and galaxies interact with one another. Because these forces are so fundamental to our existence, before considering how they work it is helpful to introduce them. They are the stars of the book, as without them you and I would not be here, and nor would the atoms from which we are made.

All objects in the universe are made up of atoms. Each atom consists of a central blob called a nucleus and electrons that whizz around the nucleus at various distances from it. Atomic nuclei are made from smaller particles called protons and neutrons. Each atom of an element has the same number of protons and electrons, and, except for one type of hydrogen that has no neutrons, each atom also contains one or more neutrons, with the number depending upon the element. Hydrogen atoms always have one proton and one electron. However, in nature, hydrogen nuclei can have either zero, one or two neutrons. These different types of hydrogen are called protium, which has no neutrons, deuterium, which has one, and tritium, which has two. These three types of hydrogen atom are known as isotopes of hydrogen. Neutrons and protons are built from quarks. The quarks are held together by a force called the strong nuclear force. This force also holds together the protons and neutrons to make atomic nuclei. The strong force prevents neutrons and protons from drifting apart, while also preventing quarks from separating from one another. If the strong nuclear force did not exist, neither would atomic nuclei, nor life.

The next force to introduce is called the weak nuclear force. It also operates within atomic nuclei, and it is the reason that isotopes of some elements are radioactive. Most atomic nuclei are stable and one of their neutrons doesn't suddenly morph into a proton, or vice versa, creating an atom of a different element. However, some isotopes of some elements have an unstable configuration of neutrons and protons, and this can cause a proton to become a neutron, or a neutron to become a proton. When this happens, a type of radiation is emitted. Not all radioactive elements decay like this, but ones that do, like tritium (the isotope of hydrogen with two neutrons), which decays into helium, do so because of the weak nuclear force. You might think that the weak nuclear force has little to do with daily life, but without it the nuclear fusion that powers our sun could not happen. Nuclear fusion is caused by the weak nuclear force, and in the sun it is responsible for creating helium from hydrogen isotopes, and this generates a lot of energy in the form of heat and light. Without the weak nuclear force, our universe would be dark. No weak force, no life.

Protons and neutrons are often depicted in diagrams as tiny balls that stick together to form atomic nuclei, and that is a reasonable analogy, even if it is not quite accurate. A blur, or fuzz, of electrons whizzes around the nuclei in a type of orbit that chemists call an orbital. The electromagnetic force keeps the electrons in place. If electromagnetism did not exist, electrons would drift away from atomic nuclei and atoms could not exist.

Electromagnetism does more than this. Atoms come in different types, determined by the number of protons, neutrons and electrons. The number of protons and electrons in an atom is equal, and this number determines the element. For example, lithium has three protons and three electrons, while carbon has six of each. When a chemical reaction occurs, atoms share or exchange electrons with one another thanks to the electromagnetic force. When they do this, they form molecules. Water is an example of a common compound made of molecules. Each water molecule consists of two atoms of hydrogen and one atom of oxygen bound together by the way they share electrons via the electromagnetic force. Carbon dioxide, methane and rust are examples of other molecules. Molecules occur in an extraordinarily diverse array of forms. but each is made of atoms, and each atom is made of protons, neutrons and electrons. No electromagnetic force and no atoms or molecules.

The final force to introduce is gravity. We rarely think about the fact that we can walk around on Earth without floating off into space. Gravity is responsible for anchoring us to our planet. It is a force that attracts objects towards one another, and it means that things with mass, including atoms, have a tendency to clump together. Gravity impacts large objects as well as small ones. It keeps the moon orbiting the Earth, the Earth orbiting the sun, and the sun orbiting the centre of our galaxy. You will know that it takes one year for the Earth to go around the sun. It takes 27 days for the moon to complete an orbit of Earth, and about 250 million years for our sun and solar system to journey around the centre of the Milky Way. We are on a rock orbiting a star that is orbiting the centre of our galaxy. Gravity is the force that keeps this dance going and, without gravity, life would not exist, as there would be no sun, no Earth and no moon.

These four forces are all central to life in that they determine how our universe behaves. Understanding the forces is consequently an important step in the narrative of why we exist. We are made out of particles that have combined to make atoms and molecules that in turn have combined to form you and me. Our next step is to investigate how the four fundamental forces interact with, and impact, matter.

Particles can combine and interact in multiple different ways. The rules of how they interact are constant across the universe, but the outcome of the interactions depends upon the way that matter is spread out across space. Part of the reason for this is that some forces operate across very short distances, while others operate over much larger ones. For example, the strong nuclear force operates at the smallest scale, followed by the weak nuclear force, with electromagnetism and gravity working up to very large scales.

The way that the strong and weak nuclear forces and electromagnetism work is very similar, with quarks and electrons continually exchanging things called force carrier particles with one another. The force carrier particles for the strong and weak nuclear forces are respectively referred to as gluons and bosons, while the force carrier particle for the electromagnetic force is called the photon. Scientists suspect gravity works via a force carrier particle too, but they have been unable to detect it. They call the hypothetical particle the graviton. Detecting gravitons would require more energy than is available to us on Earth. Some estimates suggest that we would need to build a machine like the LHC that was the size of our solar system, that we would need to position it next to a very dense star, and that even then we would observe only one graviton every decade or so. If they exist, gravitons are elusive. Fortunately, force carrier particles that determine the other forces are easier to detect, and they have all been observed. Each force carrier particle is associated with a force field, with each force having its own field.

These force fields permeate the universe in every direction and at every point, even in deep space between galaxies. The easiest way to imagine these fields is by considering the force field around a magnet. We can see this magnetic field by sprinkling iron filings on top of a sheet of paper placed above a magnet and gently tapping the paper. The iron filings trace arcs from one pole of the magnet to the other. What you see when you do this is a trace of the electromagnetic force field. The other force fields are not as straightforward to observe, but they are everywhere all the time.

Each force field consists of energy, and there are continuous fluctuations in this energy that result in the force carrier particles appearing and disappearing. They do not exist for long, only a fraction of an instant of a second, before disappearing again. Force carrier particles are sometimes referred to as virtual particles, although they are, in fact, very real. These virtual particles pop in and out of existence more frequently close to matter than further away, but despite this, even in parts of the universe a long way from any matter, energy is morphing into these particles and back again.

Although these virtual particles are small and cannot be broken into smaller components, there may be things happening on an even smaller scale. Some physicists argue for a theory, called string theory, that hypothesizes that there are tiny, vibrating, string-like structures that determine all of matter. The frequency at which each string vibrates determines the type of particle we observe. At the scale of strings, space is thought to look very different to the way we perceive it. We think of space as being smooth, and not being divided up into discrete chunks. As we walk or drive, we do not jump from one location to a neighbouring one, but rather we move through space smoothly. On the extremely tiny scale where vibrating, string-like structures are thought to exist, scientists believe that space loses this continuity and becomes divided up into minuscule volumes, a little like an infinite stack of miniature square boxes. Physicists refer to this scale as the Planck scale, and the Planck length (the side of one of these boxes) is the smallest length that exists. It cannot be sub-divided. The idea of half a Planck length makes no sense. At the Planck scale, tiny strings might be vibrating, and this would make space look like bubbling foam if we could observe it at such a minuscule scale. These conclusions are reached through the analysis of mathematical models because we do not have the technology to see space at the Planck scale.

The Planck scale is not the only aspect of the universe that remains elusive to physicists. Our universe consists of matter (including something called antimatter; more on that later), in the form of fundamental particles; energy, in the form of heat and light; something called dark energy, which must exist but that scientists are yet to fully understand; and virtual particles that are continually being produced and destroyed that allow other particles to interact. Each of these building blocks of the universe must be linked, but quite how dark energy links in is currently unclear. Nonetheless, physicists know how matter is related to non-dark energy in the form of heat and light via the most famous equation in science: Einstein's $E = mc^2$. The E stands for energy, the m for mass and the c for the speed of light in a vacuum, where a vacuum is space containing no matter. The equation links energy to matter, showing how they are related. The amount of energy tied up in matter is very large, because the speed of light in a vacuum (c) is a big number – 299,792,458 metres per second. The square of this number, 299,792,458 × 299,792,458, is an even bigger number. What this means is that if matter can be coerced into releasing this energy, as happens inside stars and nuclear bombs, a lot can be produced. Conversely, when matter is created from energy, it locks up a lot of energy.

At its birth, 13.77 billion years ago, the universe was a point of hot energy before some of this energy transformed into quarks and electrons that did not then instantly morph back into energy as virtual particles do. Instead, they survived to allow us to exist. They stuck as matter, and scientists are not entirely sure why. What they do know is that these fundamental particles soon started to interact with one another, and when the universe had cooled to a temperature of 1 billion degrees Celsius, the quarks started to join to form the protons and neutrons found in the nuclei of atoms.

Neutrons and protons are collectively known as nucleons. Each nucleon consists of three quarks, with the quarks that form nuclei coming in two different types called up and down. Protons consist of two up quarks and one down quark, while neutrons consist of two down quarks and one up quark. These quarks are bound together, or constrained, into a small space by virtual particles called gluons, with each pair of quarks continually exchanging gluons with one another. As quarks are moved apart, the strength of the containment increases as more virtual particles appear and disappear between the quarks, pulling them back towards one another. Each gluon that is exchanged exists for only a tiny fraction of a second as it moves from one quark to the other, but this movement strongly binds the quarks together.

Each quark has several characteristics that scientists use to describe them. Some of these characteristics are colour, mass, spin and electric charge. An up quark has a 2/3 positive electric charge, while a down quark has a 1/3 negative electric charge. When they combine to form protons, the two 2/3 positive charges and the 1/3 negative charge add up to give a positive charge of 1. In a neutron, the 2/3 positive charge and two 1/3 negative charges combine to give an electric charge of 0. Colour charge is conceptually similar to electric charge, but comes in three types, red, green and blue, rather than the positives and negatives of electric charge. These colours should not be taken literally, they are simply a tool that physicists use to distinguish the three types of colour charge. When quarks interact by exchanging gluons, they also change their colour charge.

The strong nuclear force not only binds quarks together to form nucleons, it also binds protons and neutrons together to form the nuclei of atoms. Two particles with the same electric charge repel one another via the electromagnetic force, much as magnetic poles of the same type push against one another. Because all protons have a positive charge, the electromagnetic force acts to push them away from one another. The strong nuclear force is sufficiently powerful to prevent the positively charged protons being pushed apart by the electromagnetic force, and it does this by binding protons and neutrons together. It does this binding not with gluons but with another virtual particle called a meson. For reasons I won't elaborate on, mesons are not defined as force carrier particles although they do play a key role in the way the strong nuclear force operates. Nonetheless, like gluons, mesons operate on only a very small scale that is slightly larger than a nucleon. Their binding ability is impressive. The binding strength of mesons is sufficient to keep multiple positively charged protons together within the nuclei of atoms, bound up with neutrally charged neutrons.

The weak nuclear force is weaker than the strong nuclear force, and much like the strong nuclear force it operates on a small scale. As with the other forces, there is a weak nuclear force field that pervades the universe, and the force is carried by force carrier particles called W and Z bosons. There are in fact two types of W boson, but we will not consider the details here.

When the weak nuclear force turns protons into neutrons and vice versa a particle called a beta particle is emitted. Beta particles are very high-energy particles and this makes them radioactive. The elements from which we are made, including hydrogen, oxygen, carbon, nitrogen and calcium, do not decay into other elements. They are stable, which is probably just as well, because if they did decay, we would quickly fall apart.

The first atomic nuclei to form in the universe were those of hydrogen and helium. These nuclei persisted in a hot plasma, along with unbound electrons, particles that have an electric charge of -1, until the universe had cooled sufficiently for the first atoms to form. At this point electromagnetism began to play a major role in the development of matter.

Electromagnetism is a hundred times weaker than the strong nuclear force, but it operates over much greater distances than either of the nuclear forces. Along with gravity, it is the force that we experience most in our daily lives. It is the force that makes compounds solid, liquid or gas, and it is the force that allows atoms to bind together in the form of molecules. Much as there are force fields throughout the universe for the strong and weak nuclear force, with gluons and W and Z bosons being the particles that carry the forces, there is also an electromagnetic force field. The particles that carry the electromagnetic force, and which allow protons and electrons to interact with one another, are called photons. These are also the particles from which light is made. Quarks exchange gluons when they interact, while electrons and protons exchange virtual photons when they interact. Virtual photons are popping in and out of existence as protons in atomic nuclei and electrons in orbitals around these nuclei interact.

The weakest force of all is gravity. It is 10 raised to the 38th power weaker than the strong nuclear force. Despite being so weak, when objects get large the pull of gravity is strong. The Earth is large enough to stop us floating off into space. Unlike the strong and weak nuclear forces, but in a similar manner to electromagnetism, gravity can operate over very large distances. Gravity and electromagnetism impact our daily lives more than the other two forces because we are very large compared to the scale at which the nuclear forces operate.

Up quarks, down quarks and electrons are not the only fundamental particles. My discussion so far has focused on only one of three groups of fundamental particles. Physicists refer to these groups as generations. I described the first generation, and that is the one all the matter we observe today is made from. I also omitted to mention a particle called the neutrino when describing the first generation. Neutrinos are extremely abundant, but they rarely interact with atomic nuclei or electrons. Perhaps somewhat obviously, you can only detect something if it interacts with something else, and not all particles are affected by all four forces. Neutrinos only interact via the weak force and gravity; they are unaffected by the strong force and electromagnetism. Every second, millions of neutrinos emitted from the sun pass through the Earth without interacting with any of the matter from which it is made. Nonetheless, a very tiny proportion of the neutrinos do interact with atomic nuclei. I did not discuss neutrinos above because they are not part of us, but they do need mentioning now to help appreciate how remarkable is the understanding of the universe that physicists and cosmologists have developed.

The first generation of particles contains four particles: up and down quarks, electrons and neutrinos (or, more accurately, electron neutrinos). Heavier versions of these particles make up the second generation, while even heavier counterparts constitute the third. As an example, the second generation of matter is made up of charm and strange quarks, the muon (rather than the electron) and the muon

Table of the Fundamental Particles

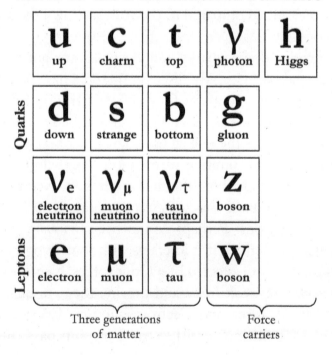

neutrino. 'Charm' and 'strange' are names rather than meaning that one of these particles is alluring and the other peculiar. The strange quark is heavier than the down quark, its sister particle in the first generation. Particles in the second and third generations of matter are not stable, and when they are formed at high energies they quickly decay to create the particles in the first generation.

The four particles in each of the three generations combine to make a total of twelve fundamental particles that are the building blocks of matter. I have also described the four force carrier particles, the gluon, photon and W and Z bosons, which allow the other twelve particles to interact. Ignoring gravitons, which are hypothesized to exist but not observed, we are up to sixteen particles. There is one more particle to introduce, and it is called the Higgs boson, taking the total number of particles to seventeen. Without the Higgs boson, no particles would have mass, and stars, planets and you and I would not exist. The story of the discovery of the Higgs boson reveals just how well physicists understand the dynamics of the three generations of matter and the force carrier particles.

The Standard Model is a theory developed by physicists to describe how the seventeen particles listed above interact and behave. The full equation of the Standard Model is mathematically daunting. It cannot be described as a particularly elegant-looking equation, but it is an amazing achievement and provides astonishingly accurate predictions on how particles interact via electromagnetism and the strong and weak nuclear forces. Although complicated, the Standard Model can be broken down into four parts. The first part describes the four forces, while the second captures how the four forces act on each of the fundamental particles in the three generations

of matter. The third part explains that the Higgs boson determines the mass of each of the fundamental particles, while the
fourth part describes how the Higgs boson goes about doing
this. The problem was, up until quite recently, no one had ever
detected a Higgs boson. The Standard Model depended upon
its existence, but what if it didn't exist? Physicists would have to
go back to the drawing board and derive a new model as their
understanding of the universe would be flawed.

Detecting the Higgs boson was never going to be easy. It
would require creating conditions similar to those of the
early universe, and that requires a huge amount of energy.
The boson would need to be detected very rapidly after it
was coaxed into appearing before it decayed into other particles. Physicists built the world's largest machine to search for
the Higgs boson, a machine I have been fortunate enough
to visit.

My eldest daughter has always been an avid reader and,
like me, she went through a spell when she devoured books
on physics. Like father, like daughter. The curriculum she
was taught had thankfully been significantly reworked since
I was at school, and she liked physics so much she went on
to study it at university, and then to work for a physics start-
up company in Oxford, the town where my wife Sonya and
I live. As Sophie's eighteenth birthday drew near, I asked her
what she wanted as a present, and she said she'd like to visit
the LHC, the machine that physicists built to detect the
Higgs boson.

One of the great joys of working at a university is you get
to meet experts in all sorts of subjects, and I happened to
know a physicist who splits his time between Oxford and
CERN. I asked him when was a good time to visit, and he
told me to come when it was shut down for maintenance.

I looked confused, and he said that if we came then, he could arrange for us to go underground and see the heart of the machine. Sophie and I duly headed to Geneva, where my friend took us to visit part of the machine called the ATLAS detector. As a field biologist who spends a fair amount of my time in wild parts of the planet, I was used to colleagues in other subjects being a little envious of where I work. I very rarely wish I had chosen a different career path, but I did suffer physics envy when we visited the LHC. It is one of science's, and humanity's, greatest achievements.

At the LHC, beams of protons, and occasionally other particles, are accelerated to very close to the speed of light before being collided with one another at one of four locations on the 27-kilometre underground particle speedway that is the core of the machine. At each of these locations there is a huge bit of complicated equipment specifically designed to detect the particles that are created from high-speed proton–proton collisions. The particles produced by these collisions do not last for very long, before they decay into other particles giving off energy, or interact with protons, neutrons or electrons. The ATLAS detector is at one of the eight locations on the LHC, and it was one of two detectors that observed the Higgs boson in 2012. The ATLAS detector measures 25 metres by 45 metres and is housed in an underground cathedral-sized cavern. The detector is designed to measure the masses, trajectories and lifetimes of the particles created when proton–proton collisions occur.

The LHC and its four detectors constitute not only the world's largest machine but also its most technologically advanced. Particles are accelerated and steered around the loop using super-cooled high-powered magnets before being crashed into one another with remarkable precision. The LHC has significantly advanced our understanding of particle physics, but the discovery of the Higgs boson has been its greatest success.

The Standard Model is almost as remarkable as the LHC. It is a mathematical formulation of how quarks, electrons and neutrinos (and higher generations of particles) interact with one another via electromagnetism and the strong and weak nuclear forces. It accurately describes what happens in the physical world around us, and it predicted the existence of a particle that had not been observed. When physicists looked for it, it was there, just as the Standard Model predicted. Thousands of physicists and engineers collaborated to design and build the world's biggest machine to detect the particle. It cost £3.8 billion to build, and costs about £1 billion each year to run, but in my opinion the cost was well worth every penny. We now have a good understanding of how the universe works. Except that not all physicists are happy. There is something missing, and this led to some scientists arguing that we need a complete rethink of the Standard Model. Any new version must perform as well as the existing model, but it will be bigger and better.

The reason the Standard Model is incomplete is it does not include gravity. It explains how the seventeen particles it describes behave in the absence of gravity. One of the fundamental forces is missing from the model. A huge challenge in physics today is linking our understanding of gravity with the Standard Model.

Physicists have a theory of gravity that is remarkable, and as impressive in its predictions as the Standard Model. The theory is called the theory of general relativity, and it was developed early in the twentieth century by Albert Einstein.

The theory of general relativity is an extension of an earlier theory that Einstein also developed called special relativity, and together the two theories describe how gravity, energy, space, time and mass are linked, and how the speed of light is central to the association between energy and mass. Einstein built on fundamental work on gravity that was begun by Isaac Newton, another colossus of science.

The universe has a maximum speed limit, which is the speed of light in a vacuum. The speed limit, as I explained above, is 299,792,498 metres per second, and this is the speed at which photons travel through empty space. If you were to approach the speed of light, some strange things would happen. First, time would slow down, and second your mass would get greater and greater. It is impossible for something with mass to reach the speed of light as its mass would become infinite, but if you were able to, time would stop. Photons can travel at the speed of light because they have no mass. Each photon is a packet of energy zipping through space.

We must exercise our imagination and forget day-to-day common sense when considering Einstein's theories of relativity. As an example, if we measure how long it takes light to travel to Earth from the sun, we will record a time of eight minutes and twenty seconds. That is our perception of time. From the perspective of the photons that constitute the beam of light travelling through the vacuum of space, each photon is simultaneously at the sun and on Earth. From the perspective of the photon, there is no time, because time has stopped. Einstein named his theory appropriately, because reality is relative to the speed at which you are travelling.

Because speed and time are linked, regardless of how fast you are travelling, if you were to switch on a torch, you would always measure light moving away from you at 299,792,498 metres per second. Regardless of whether you are standing still, or travelling at just 1 kilometre per hour slower than the speed of light, if you turned on a torch and measured how quickly the beam of light accelerated away from you, you would always record the same number: 299,792,498 metres per second. It doesn't matter how fast you are going; you would always see the same thing, light racing away from you extremely quickly. This is because one hour lasts for much longer when you are travelling close to the speed of light than when you are stationary. The speed of light is a constant regardless of how quickly you are moving.

Such behaviour is very different from our experience on Earth. If you are travelling in a car at 95 mph and another vehicle overtakes you at 100 mph, it does not appear to be travelling much faster than you. In contrast, if you are travelling at 100 mph, a car passing you at 200 mph moves away from you at a much faster speed. At these low speeds, we do not notice time running at different rates as a function of how fast we are travelling, but even at these velocities time does run at ever so slightly different speeds.

The link between speed and time has some unusual consequences. If one twin was to head off on a space journey at 99.99 per cent of the speed of light aged twenty years old, returning to Earth after what would seem to her to be one year later, she would be twenty-one years old but her twin sister would be ninety. On Earth, we think of time as a constant, always ticking away at the same rate, but this is not true when we start to move faster. Time is relative.

As I teenager I was fascinated by the thought of travelling at light speed, even though I knew it could never happen. I wondered what the universe would look like out of my spaceship's window as it hit light speed. Because time had stopped for me, but not objects travelling at less than the speed of light, would I be able to see all the future locations of all the objects in the universe? Would Earth's orbits of the sun appear like a smear across space? In imagining this, I had made an error of logic, but I learned an important lesson. My mistake was forgetting that space and time are linked. As speed increases towards the speed of light, and time slows, space contracts, and this means the universe would appear to get smaller and smaller until, at light speed, it would appear as a very bright but infinitesimally small dot. The lesson I learned is that science proceeds by trial and error. What might be a good idea may prove to be wrong. Science is about explaining patterns in data by ruling out hypotheses that turn out to be incorrect. In forgetting that space contracts as speed increases, I had arrived at a flawed hypothesis but learned something important: that science proceeds one disproved hypothesis at a time.

Einstein got it all squared away. He showed not only that space and time are changed by speed but also that they are impacted by gravity. His general theory of relativity includes space, time, speed and gravity. The stronger gravity is, the slower time flows. It is even possible to measure this effect on Earth, with time running faster at the top of a mountain than at sea level. The effect we see on Earth is very small but is exactly what Einstein's equations predict.

Gravity bends space as well as slowing time. As a beam of light travels through space, if it passes by an object with a large mass, its trajectory bends. A widely used analogy to describe the effect of gravity on space, but in two dimensions rather than three, is to imagine a trampoline with nothing on it. Let us assume that the surface of the trampoline is perfectly flat, such that if you were to roll a marble across it, it would continue in a straight line until it fell off the other side. Now repeat the experiment but having placed a weight in the centre of the trampoline. The weight is sufficient to cause the trampoline to sag in the middle. If your marble is travelling fast enough, its trajectory across the trampoline will now be curved. The gravity of an object curves space much in the same way the weight has changed the surface of the trampoline from a flat plane to being depressed in the middle. And if the heavy object is very heavy, your marble's trajectory will be so sharply curved it'll go into orbit around the heavy object.

The greater the mass of an object, the stronger gravity becomes. The sun, for example, has a much stronger gravitational pull than the Earth because it has over 333,000 times more mass. The sun bends space to a greater degree than the Earth, and time runs slower the closer to the sun you get. But even the sun's gravitational pull pales in comparison to that of black holes. Around each black hole is something called an event horizon. It is the point at which gravity becomes so strong that even light cannot escape its attraction. The gravity in black holes folds space in on itself and stops time for any object that crosses the event horizon.

Einstein's theory of relativity has repeatedly been tested and has always been found to be remarkably accurate. It describes how (and why) the moon orbits the Earth and the Earth orbits the sun, and why we do not float off into space; it has been used to age the universe and to reveal why it continues to expand. It also reveals that there must exist another type of matter. Without this matter, galaxies would not exist.

When scientists discovered galaxies spinning at such a speed that solar systems towards the edge should not be held in their orbits by the gravitational pull of the rest of the galaxy, they had a problem. The speed at which the galaxies spun meant the solar systems towards the galaxy's edge should be jettisoned into space. Cosmologists resolved this contradiction by devising a hypothesis that particles exist that do not interact with other particles via the strong and weak nuclear forces and electromagnetism. They called them 'dark matter'. According to the theory, they only interact via gravity. This makes them extraordinarily hard to detect and, so far, scientists have not been able to observe dark matter particles, even with the LHC. Nonetheless, dark matter particles must be very common given Einstein's theory appears to be correct.

Dark matter is not the only stuff that scientists hypothesize must exist if Einstein's theory is true. The universe is expanding quickly, and its rate of expansion is increasing. Physicists have calculated that something other than dark and observable matter must exist to explain this expansion. On very large scales, something is pushing galaxies apart. Scientists have termed this mysterious stuff dark energy briefly mentioned earlier in this chapter. Dark matter and dark energy must be extraordinarily abundant given the observations of the universe we have made. Estimates suggest that 5 per cent of the universe is made up of energy and matter we can see, 27 per cent is dark matter and a whopping 68 per cent is dark energy.

Although scientists know little about dark matter particles or where dark energy comes from, there are many measurements that suggest they must exist. The fact that particles and energy exist that we know so little about is not entirely bonkers. Before the discovery of the Higgs boson, you might have argued that the Standard Model must be wrong because a particle fundamental to it had not been measured. The fact we have not yet been able to directly measure dark energy and see dark matter does not mean it does not exist. In addition, as you will have gathered, not all particles in the Standard Model are influenced by all four of the fundamental forces. In fact, it is only the quarks that feel the strong and weak nuclear forces, electromagnetism and gravity. Electrons are not impacted by the strong nuclear force, and neutrinos are so hard to study because they only interact with other particles via the weak nuclear force and gravity. If dark matter, as is hypothesized, does not interact with the electromagnetic or nuclear forces, the particles that constitute it will be hard to directly detect, requiring levels of energy well beyond those generated by the LHC.

Let us briefly recap where we are at before returning to our story of the history of the early universe. Scientists have one theory, the Standard Model, that is astonishingly good at describing how three of the fundamental forces interact. They have a separate theory that explains how the fourth force, gravity, works. But they have a problem in that merging the two theories into a theory of everything is mathematically extraordinarily challenging. The Standard Model breaks down when gravity distorts space and time, and it cannot be extended to incorporate gravity using the mathematical tricks that were fundamental to its development.

Despite this, there are theories, including string theory I mentioned earlier, M-theory and quantum loop gravity, that are proposed as ways to unify gravity with the other three fundamental forces. Currently we have no way of testing the predictions each theory makes, and this means a widely accepted grand theory of everything does not exist. A general theory of everything would help us further probe the history of the early universe, but even without such a theory, there is quite a bit that scientists have been able to discover about the youthful universe. They have also been able to run computer simulations where they change the strength of the four forces and see whether atoms, and stars, would form. Not all values of the four forces would result in a universe where life could evolve, so we should be grateful they take the strengths they do. We do not know why the forces have these strengths, but I'm glad they do. If they didn't, you wouldn't be reading this book.

Insights from the Standard Model, the theory of general relativity, collisions between tiny particles in the world's most complicated machine, computer simulations, and observations of galaxies across the heavens have allowed scientists to piece together the history of the early universe. With these insights we can step from energy to quarks and electrons to atoms of a whole host of elements, the first necessary parts of the universal history of you and me. What happened is what I summarize next. The history of the universe as it developed from a singularity to the stars we see when we gaze at the night sky.

Before our universe's birth, scientists speculate there was a nothingness. When I say nothingness, you might imagine the emptiness of space, perhaps like the space between the Earth and the moon, or between the edge of our solar system and Alpha Centauri, our closest star system. But this space is not empty. It contains energy and force fields and on the Planck scale might look like a bubbling foam. Scientists assume that before the universe formed, there was nothing. Nothingness means no energy and no force fields. Somehow from this nothingness our universe materialized, and scientists do not know how. Some researchers have put forward theories, but currently there is no way of collecting data to test them. For instance, some mathematicians have argued that the nothingness from which our universe was born was unstable, and it collapsed to form the singularity that was our very early universe, but in truth we have no way of knowing whether this hypothesis is true.

The singularity that appeared 13.77 billion years ago and from which our universe grew may have been over a billion billion degrees Celsius and was very, very dense, with all the mass and energy of today's universe contained in a single point that was smaller than the smallest particle seen in the universe today. The four fundamental forces of physics that now permeate our universe did not exist at its birth, and the force (or forces) at play in the first instant of our universe may have been simpler than those that govern our lives. In a fraction of a fraction of a second after the singularity appeared, there was a brief period of what scientists refer to as inflation that lasted less than a billionth of a second, during which time the universe expanded astonishingly quickly. During this period the universe grew by 1 followed by 78 zeros times. Even after this period of rapid expansion the universe would have only been somewhere between the size of a grain of sand and a basketball. In the remainder of the first fraction of a second of the universe's life, the four fundamental forces emerged.

As the young universe expanded it began to cool, and as it did so the fundamental forces we see today began to reveal themselves. First to appear was gravity, followed by the strong nuclear force. Cosmologists do not agree whether the strong nuclear force separated from the electroweak force – the combined weak and electromagnetic forces – immediately before or after inflation, but its emergence filled the universe with a hot soup of quarks and gluons. Very shortly after the emergence of the strong force, the electroweak epoch ended, when the two remaining forces – electromagnetism and the weak nuclear force – appeared and electrons were formed. All this happened extremely quickly, and by the time the universe was 0.000001 seconds old, it had cooled sufficiently that neutrons and protons were able to form.

The hot soup of protons and electrons was opaque, and energy, in the form of photons, could not travel through it as easily as they traverse the universe today. The early universe by was consequently dark. Cosmologists study the universe by examining different types of electromagnetic radiation: gamma rays, X-rays, ultraviolet light, visible light, infrared light, microwaves and radio waves. These different types of electromagnetic radiation travel across the universe as photons that are not only particles but are also waves (I explain this in the next chapter). Electromagnetic radiation lies on a scale defined by the distance between the peak of consecutive waves. Radio waves have the longest wavelength, and gamma rays the shortest. None of them could travel through the early universe, as the photons that carry energy would continually interact with the quark–electron soup.

By the time the universe was 380,000 years old, it had cooled sufficiently that the protons, neutrons and electrons could combine and form atoms of hydrogen and helium. At this point, the universe became transparent, and electromagnetic radiation could travel through it. The oldest radiation we can detect on Earth today has been travelling through the universe since it was 380,000 years old, a journey taking over 13.7 billion years. It is called the cosmic microwave radiation.

The cosmic microwave radiation is everywhere we look, and it has very similar properties whether we look towards the centre of our galaxy, or away from it. The model that ran time backwards from now towards the singularity at the birth of our universe predicted that such radiation should exist. Discovering it helped prove that the universe started with a singularity and a big bang.

Since the 1960s, when the cosmic microwave radiation was discovered, many research projects have mapped it, with a very high-resolution all-sky map being published in 2013. The map revealed the radiation was not completely identical in every direction but instead exhibited very small variations. There are points where it is slightly warmer, and others where it is a bit cooler. Because this radiation permeated the universe from its birth, this patchy temperature variation reflects slight differences in the distribution of energy and matter in the very early universe. Scientists believe that this variation early in the universe's life is what resulted in the formation of galaxies and the large-scale structure of the universe we see today.

The detailed map of the cosmic microwave background also provided compelling support for the period of inflation, with the small fluctuations being caused by tiny differences in the distribution of matter during the universe's very early history. The map also allowed cosmologists to calculate the age of the universe to a precision of 13.772 billion years old, plus or minus 59 million years.

Because the universe is so staggeringly vast, light that arrives on Earth may have been travelling through space for millions, or even billions, of years. When we look at an object in the night sky, we do not see what is happening to it now, but rather we see what it was doing when the radiation that we see left the object. Light departing the moon takes 1.3 seconds to reach Earth. Light from Alpha Centauri takes a little over four years and three months to get to Earth. If something catastrophic were to happen to the Alpha Centauri system, we would not know about it for over fifty-two months. When we gaze into space, we are consequently looking back in time, a bit like a geologist digging down through layers of time in rocks. The geologist is looking at a still picture of the past, while the astronomer sees a movie. If alien astronomers orbiting a star 70 million light years away are at this moment pointing their immensely powerful telescopes at Earth: they are eyewitnesses to tyrannosaurs charging across the Cretaceous landscape.

The further away an object is, the longer it has taken the electromagnetic radiation to reach us. By looking at objects that are different distances away, scientists can consequently piece together the history of the universe. The observable universe has grown from the small local neighbourhood we were in when the universe became transparent. The cosmic microwave background is the first light from the bit of the universe our current neighbourhood was in when the universe became transparent. All the galaxies we can see spread out over billions of light years are the matter that was packed into a small locality when the universe was 380,000 years old.

One way to visualize this is to imagine putting a dot of ink on an uninflated balloon. The surface of an uninflated balloon is a bit like the whole universe, but in two dimensions. When the universe became transparent and light could start to travel through it, our neighbourhood was like the ink dot on the uninflated balloon. Now imagine blowing the balloon up. As the surface of the balloon stretches, the ink dot becomes bigger. Inflating the balloon mimics the universe expanding. The stretched ink dot is like what we can see now, the observable universe. There is a lot of the balloon we can't see, and that is equivalent to the unobservable universe. The cosmic microwave background can be thought of as the light that was in the ink dot on the uninflated balloon. The galaxies represent all the matter that existed in the ink-dot region of the uninflated balloon. Cosmologists know the universe is still expanding and the rate of expansion is accelerating. This is like the balloon being blown up at an ever-increasing rate. The 'universe as a balloon' analogy is useful, but not perfect. The surface of an expanding balloon is curved, while scientists have shown that the universe is not curved but flat. The balloon analogy has other limitations too. Over-inflated balloons pop. Let's hope the universe doesn't eventually do the same.

By the time the universe became transparent it had expanded quite a lot from the singularity it started out as and was about 1.4 million light years across. The reason we cannot see the edge of the universe is that the observable universe we see today was not on the edge of the universe when it became transparent. When the universe was 380,000 years old the bit that became the universe we observe today was surrounded by matter and energy in all directions. We have no idea whether we are closer to the edge or the centre of the universe, and perhaps we will never know.

Once the universe had become transparent and hydrogen and helium atoms were common, gravity then began to work its magic. The first atoms and molecules were attracted together until they formed the first stars. These stars first burned when the universe was about 100 million years old, and they were many times larger than the sun, but their lifespans would have been shorter.

Stars burn bright because the force of gravity pulls atoms of hydrogen and helium closer together, heating them up and making them energetic enough that they start to collide. These collisions can break the bonds between nuclei and electrons, eventually forcing the nuclei together, where they can form the nuclei of heavier elements such as nitrogen and oxygen.

Brighter stars use up all their hydrogen and helium fuel more quickly than those that are smaller, but all stars above a certain size turn their fuel into nuclei of heavier elements such as nitrogen, oxygen, carbon and iron. As these stars near their end, collapsing in on themselves, gravity becomes so strong it pushes numerous protons and neutrons together to form the nuclei of the even heavier elements such as uranium and plutonium. It is fascinating how atomic nuclei of heavier atoms can be formed from hydrogen via the weak nuclear force and gravity.

In the centre of a star, hydrogen nuclei are turned into helium nuclei and something called a helium core forms. The nuclear fusion in a star is triggered because gravity pushes atoms closer together, making them highly energetic. The fusion generates a lot of energy in the form of heat and light that exerts pressure, preventing the outer layers of the star collapsing in on themselves. For most of a star's life, the effects of gravity pushing atoms towards one another, and the pressure created by nuclear fusion emitting energy, are approximately in balance. Nonetheless, towards the end of a star's life things become unstable. Exactly what happens at this point depends upon the mass of the star. In the case of the sun, it will continue to get steadily hotter and larger over the next 5 to 6 billion years until a time when it will go through cycles of expansion and contraction. At its largest size it will extend to the orbit of Jupiter, before contracting again to be smaller than it is today. These pulsing contractions and expansions will lead to the helium core becoming heavily compressed and being raised to a temperature of 100 million degrees Celsius. The helium is converted into carbon and oxygen and other heavy elements, generating huge amounts of energy that cause the sun to expand to a vast size once again. Eventually these cycles of expansion and compression cease, with the heavier elements created in the sun's death and any unburned hydrogen forming a cloud of dust. The Earth will be swallowed and burned up during one of the sun's dying expansions and our planet will cease to be.

When stars that are greater than about eight times the mass of our sun run out of fuel, gravity causes them to collapse until they form supernovae – the most violent explosions in the universe. The star that goes supernova explodes and as it does so it produces elements heavier than iron. The elements, dust and debris from a supernova permeate space and can form nebulae from which the next generation of stars is made. Heavy elements are quite abundant on Earth and elsewhere in our solar system, providing evidence that our sun cannot be a first-generation star. In fact, scientists know that the universe was already over nine and a quarter billion years old when our solar system formed. The Earth may be formed from the debris of a second-, third- or fourth-generation star – we do not know. What we do know is that the heavier elements made from the death of earlier generations of stars were necessary to produce our home planet and the emergence of life.

At this point in our story we have gone from a singularity to our neighbourhood in the universe where some of the hydrogen and helium formed earlier in the universe's life has been converted, via a process known as nucleogenesis that happens in the heart of stars, to heavier elements. Our next step on the journey is to investigate how these heavier elements combine to form molecules like water, carbon dioxide and methane – groups of atoms of different elements that have combined via the electromagnetic force.

In this chapter we have covered how the following things had to happen for you, and me, to exist. 13.77 billion years ago a very hot, very dense, singularity consisting of large amounts of energy had to form. The universe began to expand and started to cool, with its expansion and cooling continuing to this day. As the universe started to age, but still within the first blink of an eye, the four fundamental forces emerged, and energy was converted to matter in the form of the fundamental particles. The link between energy and matter is critical to our universe's functioning, and to you and me. Quarks soon started to form protons and neutrons and then the nuclei of the lightest elements. As the universe continued to cool further, the nuclei combined with electrons to form the first atoms. These were eventually pulled together to form the first stars, creating local pockets of intense heat that created heavier elements, which eventually, in our neck of the cosmic woods, combined to create our sun and solar system. I wish I'd been taught this at school, as every child should know what remarkable knowledge physicists have of how and why atoms of the periodic table came to be. Chemistry is the next step in our history, and in particular why chemical reactions happen. You and I are made of complex chemistry, so understanding how and why atoms interact as they do is crucial to figuring out why we are here.

Molecules

There were only a few bits of chemistry I liked at school, and those were the bits that resulted in explosions. Such violent reactions were few and far between, mainly because my teachers were all good chemists who made sure to give us chemicals that would not blow up when mixed. I did have a chemistry set at home, but it was hard to generate much chaos with that either. One of the reasons, as I eventually learned, is that the chemicals that are most explosive need to be kept in unreactive liquids, and for this reason were not included in chemistry sets and were kept locked up at school. Metals like sodium are kept in oil because as soon as they encounter air, they start to react with the oxygen in it. However, even the reactions of the most volatile chemicals are tame compared with the death of stars described in the last chapter, or the consequences of mixing matter and antimatter. Most of this chapter will be about reactions due to the electromagnetic force, and how electrons behave, organize themselves around atomic nuclei and form various types of atomic bonds. But before we start to focus on electrons, there is one other type of interaction to briefly consider as it too is central to our existence.

Shortly after our universe's birth, when quarks, electrons and neutrinos came into existence, so too did particles of antimatter. These are called anti-quarks and positrons, and the anti-quarks can combine to form antiprotons and even antineutrons. Some of the most impressive explosions in the universe would have occurred when matter and antimatter collided, as they destroy one another and generate vast amounts of energy. For example, if you collide a negatively charged electron with its antimatter equivalent, a positively charged positron, they emit two highly energetic photons. Antimatter is rare in our universe because shortly after an antimatter particle is created it collides with a matter particle and they annihilate one another, creating energy. One unsolved mystery of science is why matter won out over antimatter early in our universe. Theory suggests that matter and antimatter should have been produced in equal amounts, and they should have instantly completely annihilated each other, leading to a very premature end to our universe. The theory must be lacking in some way because there was ever so slightly more matter than antimatter, and that led to our universe existing as it does today. If antimatter and matter had been produced in equal amounts, we would not exist.

Antimatter was not only formed at the birth of the universe. It is produced, sometimes in the most unexpected of places. The bananas in your fruit bowl create a steady stream of positrons, with each banana producing one positron every hour and fifteen minutes or so. Bananas contain an element called potassium, including a rare isotope of the metal named potassium-40, which very, very slowly decays via the weak nuclear force to calcium-40. If you had a gram of potassium-40, just under half of it would decay to calcium-40 in a billion years. We cannot predict when any particular atom of potassium-40 will decay via the weak nuclear force to calcium-40, but we know that on average it takes 1,290,000,000 years. Because this is an average, some potassium-40 atoms will decay much more quickly, while others may last for several billion years before decaying. Given that one atom of potassium-40 decays in each banana approximately every seventy-five minutes, this suggests that there must be a lot of potassium-40 atoms in each banana. There are. Just one gram of potassium-40 contains 15 followed by 21 zeros atoms. An average banana contains only about 0.015 g of potassium-40, but that is still a very large number of atoms. When a potassium-40 atom decays it releases a beta particle – in this case a highly energetic positron – which is annihilated when it encounters an electron. Our eyesight is not good enough to see the annihilation, but it does occur, and it releases energy. If we were to scale things up by annihilating 1 gram of matter with 1 gram of antimatter, it would produce the energy equivalent of the atomic bomb dropped on Hiroshima. We are fortunate that antimatter is not produced too rapidly in our fruit bowls. If it was, making a banana smoothie could be lethal. Antimatter plays no further role in the story of why you and I exist, but be thankful it is no longer very common in our universe. If it was, you and I would not be here.

Chemical reactions do not involve the annihilation of matter and antimatter. Instead of being driven by the weak nuclear force, they are determined by electromagnetism. Despite this, some chemical reactions can still be astonishingly violent. 1-diazidocarbamoyl-5-azidotetrazole, or azidoazide azide as it is sometimes called, is one of the most volatile molecules known to humanity. It is close to impossible to store, as a small change in pressure, exposure to light or even a tiny shift in temperature can cause it to explode. It is so unstable that standard instruments used to measure how volatile a chemical is cannot be used to measure it, as attempts to use them make it explode. At the other end of the spectrum, there are molecules that are very difficult to engage in chemical reactions. Helium, for example, is very stable and tends not to get involved in reactions with other substances. It consists of two protons, two neutrons and two electrons. This does not mean that helium is not useful; it has many uses. It is lighter than air such that helium balloons float, and it is also amusing to take a breath of helium and experience your voice jump up in register. What is it that makes some chemicals more reactive than others?

In the previous chapter I introduced the four fundamental forces but focused most on the nuclear forces. I described how the life cycle of stars fused hydrogen nuclei into helium nuclei, and helium nuclei into those of heavier elements. These nuclei then joined with electrons to create atoms of the heavier elements. The next key part of our history is the advent of molecules caused by elements joining together to form a huge array of chemicals. They do this via the electro-magnetic force, and it is time for this force to take centre stage. The force drives all chemical reactions, from simple interactions between small molecules to the way that some drugs stop some complex molecules from doing their jobs inside your cells.

Although I didn't particularly enjoy chemistry at school, it did play an important role in my early life. While I was grow-ing up, my parents were drug dealers. They ran a chain of chemist shops in Cambridgeshire, dispensing prescriptions and selling over-the-counter treatments for minor ailments, illnesses and injuries. I did briefly toy with the idea of going into the family business, but I'd never enjoyed helping out in the shops as a teenager. It wouldn't have been the right job for me. Given my parents were pharmacists, they were prob-ably a little horrified by my chemistry reports from school. Nonetheless, as with my extracurricular physics reading, I spent time at home reading around the subject, but not in quite the same way as I did with physics. I failed to develop a good grasp of electromagnetism, but I did learn how molecules could treat and prevent disease. My parents had a large book listing endless ailments and diseases and the drugs that could be used to treat them. It was fascinating reading, and remarkable to learn how many different ways there were to become sick, and ways in which drugs can relieve symptoms. The problem was that every time I got a temperature, spot or rash, I would self-diagnose, always fearing the worst. Being pharmacists, my parents were very responsible with drugs and, apart from the occasional aspirin, my requests for drugs to treat scurvy, Guinea worm, elephantiasis or the bubonic plague fell on deaf ears.

At about the age of ten or eleven I read about a disease that put the fear of God in me because it had no cure and a terrible end: rabies. Each year we would head to France on holidays, and for a couple of these trips I hardly slept. I was convinced every cat, dog, cow or sheep I saw was rabid and out to get me. I imagine that my intense anxiety meant Mum, Dad and my sister, Fiona, didn't enjoy the trips much either.

I did at one point think I might have caught rabies. During the time I spent in Zimbabwe in my late teens, I did get bitten by a dog, and for a brief spell my fear of rabies returned. I went to the local clinic and learned that I could get vaccinated, but it would require seven unpleasant injections into the stomach that could only be administered in the capital city, Harare, which was not an easy journey from where I was based. The dog bit me because it had been asleep and I had woken it up by stumbling, a little worse for wear, out of a bar and trodden on it. I also discovered there had been no cases of rabies in the area for months, if not years. I decided the risk of death-by-rabies was low and I wouldn't get the jabs.

Nonetheless, I did spend the next few weeks trying to assess whether I was turning mad or starting to develop a raging thirst. I did read up about rabies, and when I learned of a reliable vaccine being available I made sure I was vaccinated ahead of work trips that might result in my coming into contact with wild animals.

My fixation with disease, vaccines and prescription drugs helped me learn a little about just how remarkable the behaviour of some molecules can be, and I continue to be impressed with the breakthroughs chemists make. For example, the mRNA vaccines, such as the Pfizer jab that was developed during the COVID pandemic, are based on a remarkable understanding of the chemistry of our immune systems. Artificial chemicals pervade our lives, from curing sickness and preserving food to killing pests and dyeing our clothes, and they all operate via the electromagnetic force. It has been estimated that as many as 10 million new chemical compounds are synthesized by chemists globally each year, with 2,000 of these being approved annually for commercial use in the US alone. Elements can combine in a vast array of ways to form an infinite number of substances. Some of them are beneficial to us, others harmful. The chemical weapon carbonyl chloride which was used in both world wars is one of the most poisonous synthetically made chemicals. As little as 200 parts per million in the air can kill a person. That's bad, but not as bad as the botulinum toxin produced naturally by the bacterium *Clostridium botulinum*. Just 1 gram has the potential to kill as many as a million people. Some people have tiny amounts of it injected into their bodies, as it is a key ingredient of Botox.

Chemistry works because atoms of different elements join together to form molecules, and different molecules can combine in a huge variety of ways. You and I consist of vast numbers of molecules, some simple such as water, which is a small molecule, to other more complex ones such as DNA, which is thousands of times larger. If elements could not join to form molecules the universe would be a much less interesting place, and we could not exist. The next bit of our history involves investigating why elements join together to form molecules, and why some of these molecules, such as DNA, are very stable, while others, such as azidoazide azide, are less so. The periodic table helps us understand how and why elements differ.

Chemists organize the elements in the periodic table. The table orders them starting with hydrogen, the lightest element, made of atoms containing a single proton and electron, at the top left and working through the elements in increasing order, row by row, all the way to oganesson, whose atoms have 118 protons and 118 electrons. Rows and columns of the table organize elements by the way their electrons are arranged, something I describe in detail below. The last and heaviest element in the table, oganesson, does not occur naturally as it is highly unstable and quickly decays, but chemists have created five, and possibly six, isotopes of the element in the lab. It is one of twenty-four human-produced elements. I find it amazing that chemists can create elements in the laboratory that are not found in nature. Is Earth the only planet where these elements have been created, or are there aliens elsewhere in the cosmos who have also made oganesson?

The periodic table separates elements that are metals, such as lithium and sodium to the left, from non-metals like chlorine and oxygen to the right. Each column groups elements by their properties, such as how they interact with other elements. Metals on the table want to donate electrons when they react with other atoms, while non-metal elements to the right want to receive them. In contrast, those in the middle of the table are more likely to share electrons. Chemists now have a good understanding of why elements behave as they do, as well as the number of protons, neutrons and electrons different atoms of different elements have, but it took a long time to achieve these insights. Exactly what constituted an element was unclear for thousands of years. Answering the question of what things are made from is hard.

People have been interested in elements that make up nature since the time of ancient Greece, and probably even earlier. Nearly 2,500 years ago Empedocles claimed that there were four fundamental elements: earth, fire, air and water. Half a century later Democritus of ancient Greece and Kamala of India independently argued that matter could be broken down into indivisible parts, but it was unclear what these parts were. The Greeks referred to them as atoms but realized they were too small to be seen.

Throughout the Middle Ages, the focus on chemistry turned from identifying elements to making useful substances, and in particular to the possibility of turning cheap and abundant metals like iron into rarer and more expensive elements such as gold. Observation had shown that new compounds could be made by combining specific ingredients, and a primary goal of alchemy was to turn one metal into another. Alchemy remained a respectable science through Isaac Newton's lifetime and was one of his interests. These days we think of alchemy as quackery, but it was of major interest to one of humanity's greatest thinkers, one of the two people who gave us calculus, a key dialect of the mathematical language of science.

By the eighteenth century, scientists and philosophers

were beginning to turn their backs on alchemy, and the era of modern chemistry began. Throughout the 1700s, various elements and compounds, both gases and metals, were isolated and named. Hydrogen was isolated in 1766 and named 'inflammable air' by the British scientist Henry Cavendish. The Swedish chemist Carl Wilhelm Scheele discovered oxygen in 1773, with it being independently discovered a year later in England.

The first chemistry text describing elements was published by the French chemist Antoine Lavoisier in 1789 and it listed hydrogen, nitrogen, oxygen, phosphorus, sulphur, zinc and mercury as substances that could be broken down no further. Other chemicals that we now know are not elements were also listed, but despite these errors Lavoisier's book was an important summary of chemical knowledge and theory towards the end of the eighteenth century. In the early nineteenth century new elements were discovered at a faster rate, with the English chemist Humphry Davy discovering sodium, potassium, calcium, boron, magnesium, strontium and barium primarily through using a technique known as electrolysis, which uses electricity to break down a compound into its constituent elements.

By the late 1860s, nearly a century after the discovery of hydrogen, sixty-six elements were known, and had their weights estimated. Mendeleev, a Russian chemist, arranged them into the first version of the periodic table in 1869. He revised his table in 1871, and in so doing revised the weights of some elements and predicted the existence of three elements today known as scandium, gallium and germanium. The discovery of gallium and scandium in the late 1870s, and of germanium in 1886, led to Mendeleev's periodic table becoming widely accepted. Chemistry had become a predictive science, and Mendeleev's legacy as one of the greatest chemists of all time was guaranteed.

At the same time, Mendeleev was developing his periodic table, the first theories of how elements combined to make compounds were proposed, and work began in trying to understand the way atoms joined with one another via chemical bonds. The electron was discovered by J. J. Thomson in 1897, with key breakthroughs in radioactivity made by the Polish chemist Marie Curie and the New Zealand physicist Ernest Rutherford. Quantum mechanics, of which more soon, was formulated in the first half of the twentieth century, and by its second half, chemistry had become a mature science. Humanity had a good understanding of how and why chemical reactions worked, and they could even predict many of them.

The development of chemistry is a fabulous example of the application of the scientific method, but it also reveals that progress can take generations. Early observations were that some substances could react, for example with iron and silver seen to oxidize when exposed to water and air, and others could be broken down into constituent parts. The hypothesis that many things could be broken down into elements became refined, and experimentalists were able to identify many of the different types of elements. Mendeleev realized that many elements behaved in similar manners, and he was able to discover a way of grouping them together. This grouping suggested some underlying principles of the ways elements interacted, and chemists and physicists developed hypotheses of how these might work. Some of these hypotheses, such as the existence of phlogiston, an element found in all compounds that could burn, or the plum-pudding structure of the atom, where electrons were embedded in a soft, spongy, cake-like nucleus similar to a Christmas pudding, were proved wrong. But other hypotheses were supported, and ever more elegant, and technologically challenging, experiments were designed and conducted.

As with physics, mathematical equations were developed to capture chemistry and the workings of atoms and molecules. The electromagnetic force became well understood, and this allowed it to eventually be linked to the strong and weak nuclear forces in the Standard Model. Physics is arguably the most advanced of the sciences, with chemistry following a close second. There are still open questions in both subjects, and new breakthroughs and applications will continue, but major breakthroughs will likely happen at a slower pace than in the sciences that focus on life. We will discover new uses for molecules, and new molecules to solve humanity's problems, and we will continue to unearth new nuances in the way that electromagnetism works, and hopefully link it to gravity as part of a theory of everything. But that's for the future. Let's now return to what we currently know, describing why atoms interact to form molecules that are essential for life.

Protons and neutrons, regardless of which element they form, are all made up of up and down quarks, so the nuclei of atoms of every element are made up of the same fundamental particles. Protons always have a positive electric charge, while the electrons that make the outer parts of atoms always have a negative electric charge. The difference in charge between protons and electrons means they are attracted towards one another. Despite each element being made up of the same type of positive and negatively charged particles, each element is different. It is difficult to get atoms of some elements, such as argon, to combine with other atoms via the electromagnetic force. In contrast, atoms of other elements, such as oxygen, are highly reactive, and will interact with atoms of other elements to form all sorts of compounds.

Not only do atoms of different elements behave differently when they encounter other atoms, but the physical properties of the elements also differ. Hydrogen and oxygen are gases at room temperature, mercury is a liquid, while lithium and iron are solid. When hydrogen and oxygen react to create water, the two gases combine to form a substance that is liquid. Unless, that is, the temperature is below zero degrees Celsius, in which case solid ice is formed, or above 100 degrees Celsius, when steam forms. These temperatures change with pressure, which is why at higher altitudes, where the air pressure is lower, steam is formed at temperatures lower than 100 degrees Celsius.

You are a complex form of chemistry, and to understand why you came into existence it is necessary to understand some chemistry. Millions of chemical reactions occur within each of your cells every second, and these reactions require atoms of elements that are reactive. If all elements were inert, like helium and neon, life could never have got started. The next step in our journey from the Big Bang to you is from atoms to molecules. If you found any of the concepts in the previous chapter strange, such that the universe has a maximum speed limit, or that gravity can slow time, or that the universe would look like a bubbling foam if it could be observed at the Planck scale, things continue to be peculiar at the quantum scale at which atoms share or exchange electrons.

In the world we experience, things do not suddenly disappear from one place and reappear in another. My dog,

Woofler, does not suddenly vanish from the bedroom before instantaneously appearing in the garden, however much he might like to do that. Similarly, objects can only be in one place at a time. I cannot simultaneously be in the office and down the pub, however hard I try. The world of the very small inhabited by fundamental particles and atoms is rather different.

All electromagnetic radiation, such as visible light, travels in waves. Yet such radiation is made up of particles called photons. The world of the very small is strange because photons, and indeed other particles such as electrons, individual atoms and even small molecules, are simultaneously both particles and waves. Being two things at once doesn't make much sense given our experience of always only being one thing: a body. However, believe it or not, you, the trees in the countryside, the cars on the road, the planets, and indeed the entire universe, are also both solid objects and waves. The waves these big objects form are billions of times smaller than the objects themselves and they do not influence our everyday lives, so we can ignore them. Things get peculiar when an object's wave is bigger than the object itself, and this is what happens to tiny objects like photons, electrons and atoms. Objects with waves that are larger than they are exhibit what scientists refer to as quantum behaviour. One way to demonstrate this, a phenomenon also called wave-particle duality, is with a neat experiment.

Scientists have conducted the two-slit experiment numerous times, and they always get the same result: atoms can appear to simultaneously be in two places at once. To do the experiment you first need a device to fire individual atoms towards a screen that can record where they collide with it. The device and the screen are in fixed positions. Next, take a sheet of material with two, close together, parallel vertical slits cut into it and place it between the gun and the screen. The slits are wide enough for the atoms to pass through them. You now fire individual atoms at the sheet and record where they appear on the screen.

Let us start by envisaging what we would expect given the classical world in which we live. Imagine scaling this experiment up so you are firing tennis balls from a gun through slits in a sheet and on to a Velcro screen. Each ball will go through one slit or the other before sticking to the Velcro. You end up with two vertical lines of tennis balls on the Velcro screen that are determined by the positions of the gun and each of the slits.

In the quantum world, some of the atoms you fire appear to behave like scaled-down tennis balls. They will look as if they leave the end of the gun, go through one slit or the other in the sheet and will make a mark on the screen used to detect them. The mark, a slit and the gun will all lie on the same line. It turns out it is just luck that a few atoms behave like this, and most atoms will not behave like the tennis balls. Instead, other vertical lines of collisions between atoms and the screen will appear. A set of many parallel lines, separated by areas where no atoms are recorded, appears on the screen. Which slit did the atom go through when the screen records a mark that is not on a straight line with the gun or either of the slits? The answer is both: each atom goes through both of the slits at the same time. That is a truly mind-bending conclusion given our existence in the classical world where a tennis ball can go through only one slit or the other.

Each atom, and indeed any object, has waves associated with it. Each wave can be described with a mathematical expression called a wave function. The wave function for any object's location describes all possible places where the object could be, along with the probability that the particle is in that part of the wave. For large objects with very small waves, the object will always appear in the same place. For small objects with large waves, the object can have a large number of places where it could be. For an atom, the wave can be viewed as describing all possible futures if it were to be forced to reveal its whereabouts as a particle. Objects can appear to be in two places when the wave function is bigger than the particle. When I mention a fuzz of electrons around an atomic nucleus, I am describing the wave function of the electrons.

In the film *Men in Black 3*, Michael Stuhlbarg plays a character called Griffin, an alien from the planet Archanan. These aliens are described as multidimensional beings who can see an infinite number of alternate timelines and possible futures. They are unable to predict which specific future will occur, but as particular events are realized some possible futures disappear while new ones emerge. Griffin uses this ability to steer the men in black to the only future that saves the world, and while doing so he admits that the ability to see so many options at once is a bit of a burden. I imagine it would be. Griffin's power is a bit like seeing wave functions for the world around him. The wave functions he sees change as time passes, and so too do the possible futures, with some becoming more likely and others less so. A particle's wave is a little bit like that. However, things in the quantum realm are even more bewildering.

There are various things you can measure about an atom, or an electron, or other particles, including their location and how fast they are moving. Each of these properties has a wave function associated with it, and the various wave functions are linked. What this means is that the more you find out about where, say, a particle is likely to be, the less you know about its speed. If you know where a particle is, you have no information about its velocity; and if you know how fast it is travelling, you have no idea where it is. The inability to know both of these things at once is called Heisenberg's uncertainty principle, after the German physicist who first described it. The quantum world is based on uncertainties that scientists can describe with probabilities. The quantum behaviour of tiny particles and atoms is probabilistic, and ultimately it all boils down to their wave functions being larger than they are.

The wave function that describes where the atom could appear in the two-slit experiment creates what is known as an interference pattern, and it is this that creates the multiple vertical lines on the screen separated by areas where no atoms hit. One way to envisage an interference pattern is to think about the behaviour of water. Imagine a wave of water flowing towards two slits. It hits the slits and emerges on the other side. Semicircular expanding waves emanate from each slit. Each of these waves will move away from the slit before the waves collide with waves formed from the other slit. The collision generates an interference pattern of peaks and troughs in the watery waves. You can see something akin to this when the wakes from two boats meet. The water becomes choppy, with highs and lows. The same phenomenon happens with the wave function of the atom fired from the gun. The wave function of the atom is choppy, with highs describing where the particle is most likely to be, and lows where it is unlikely to show itself. The vertical stripes that form on the screen in the two-slit experiment are these highs, while the gaps between them are the lows. The two-slit experiment is a very clever way of using lots of atoms to reveal what an individual atom's wave function looks like.

As an atom is fired from the gun it forms a wave which passes through the slits before the particle is coerced to emerge from the wave at a particular location on the detector screen. The slits must be close together for the experiment to work, and the width of each slit must be close to the wavelength – a property of the wave function – of the particle for an interference pattern to form. When these conditions are met, the experiment is repeatable, and the effect is always observed. All this raises the question, what makes the particle emerge from its wave and appear in one location on the screen?

When the atom in the two-slit experiment hits the screen, its wave function is said to collapse, and it reveals itself at a particular location. It would be a bit like you simultaneously leaving a building via both the front and back doors before magically appearing in one place in the street outside. The wave function describes an object's wavelength. The bigger an object is, the smaller its wavelength becomes and the object appears in one location.

As an aside, large objects can be made to behave like small particles if the wavelengths of all their constituent atoms can be aligned in a particular way, but doing this is very difficult for anything larger than a single molecule. An approximate analogy for aligned wave functions is a line of metronomes ticking in perfect synchrony. In most cases, the atoms in an object are not synchronized like this but are more like a line of metronomes each doing its own independent thing. That is what happens in you. Your wavelength is about a billionth of a billionth of a billionth of a centimetre because the wave functions of your atoms are not in sync. Because you are much, much bigger than this, you behave as a solid object and we can ignore the wave function. If you were able to line up all the wave functions of each atom in your body, you would behave like a wave. You could behave like a quantum particle, leaving a building by multiple exits and appearing in the street. You are unable to line up your atoms' wave functions, and this means the two-slit experiment won't work for you, because you cannot fit through the extraordinarily narrow slits that would be required for the experiment to work. It would be like threading a camel through the eye of a very tiny needle. The two-slit experiment only works for objects that have wavelengths that are larger than they are, and these tend to be particles and atoms. For example, the wavelength of an electron is a couple of thousand times larger than the particle itself.

When an atom passes through a slit in the two-slit experiment and encounters the detector screen, it forms chemical bonds with the detector, becoming joined to the much larger object. The atom and screen have a wavelength, but because the wavelengths of the atoms in the screen are not aligned, the atom now appears in one place as part of the screen. If the atom were to be freed from the detector, it would regain its original quantum properties. As atoms join to form molecules, and molecules join to form large objects we can see, their wavelengths usually become unimportant for their behaviour, and they behave like objects we are familiar with in our everyday lives rather than like particles in the quantum world.

When I described the early universe, I discussed the cosmic microwave background, the oldest radiation in the cosmos, and how fluctuations exist in this radiation. Cosmologists think this variation dates to the very early universe, when it

The Two-Slit Experiment

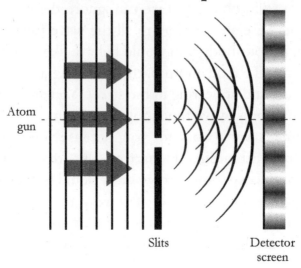

Atom
gun

Slits

Detector
screen

was so small its wave function was larger than it was. Highs and lows in this quantum wave function were amplified as the early universe rapidly expanded, resulting in variations in the distribution of matter. In time, these differences in the density of matter resulted in galaxies containing billions of stars and areas of star-free space between them. The large-scale structure of our vast universe is thought to have resulted from its quantum behaviour while it was still extraordinarily small.

The quantum behaviour of atoms is deeply peculiar, but I have not yet explained how it helps us understand the way in which atoms interact. Thinking back to my chemistry lessons at school, I was unable to find the answer to this question. We weren't taught it. In fact, I realized only three things remain with me from these lessons. First, I recall a kid called Gary managing to fill the classroom with an acrid-smelling gas that made everyone cough and led to the whole science block being evacuated. Gary spent the rest of his school chemistry career attempting to repeat the feat, but failed. Second, I remember telling my friend Mike that if he attached the Bunsen burner to the water tap it would shoot a powerful jet of water that would probably go as far as the teacher about fifteen feet away. Not believing me, he attached the Bunsen to the tap, but he didn't have the nerve to try to drench the teacher. Instead, he just turned the water tap on with the burner sitting on the bench. The water hit the ceiling and we got dripped on for the rest of the lesson. Third, we were taught that atoms are a bit like the solar system, with the nucleus at the centre, being orbited by particles called electrons that behaved like planets. There would have been better things for me to remember than the first two, and the third 'fact' we were taught is largely wrong.

One aspect of the solar system analogy that is correct is that most of an atom is empty space, but that's about as far as it goes. The simplest atom is that of hydrogen, consisting of one proton and one electron. If we were to expand the proton to be the size of the sun, the electron's most likely closest position would be ten times further away than Pluto is from the sun. That's a very long way. Mercury, the closest planet to the sun, would be closer to the nucleus than where we would expect to find any electron if the solar system were to be shrunk down to the size of an atom.

Talking in terms of probability by saying things like 'the electron's most likely closest position' sounds odd, particularly when we think of planets. It would be much easier to write that the electron is ten times further away than Pluto is from the sun, rather than that is its most likely position. Unfortunately, though, in the same way that we do not know where an atom is when it leaves the gun and before it hits the screen in the two-slit experiment, we do not know where the electron is with respect to its atomic nucleus. We just know where it is more or less likely to be, and this means we have to talk in terms of chance. Given this, we cannot describe an electron as being in orbit around an atom's nucleus. Its wave function is smeared around the nucleus and the electron can appear anywhere within this wave function if coerced to show itself. To describe this complexity, chemists use the word 'orbital' rather than 'orbit'. Electrons form orbitals around their nuclei, and there is a probability that they are at any point in their orbital. If planets formed orbitals rather than orbits, they would appear as a mist smeared across the night sky. Some bits of the mist would be thicker than others, with the thick bits denoting where the planet would most likely be. Probability adds complexity, but we do all sometimes think in terms of probability in our everyday lives.

Mine is not a family that tracks each other on our mobile phones, so as I write this in my office I cannot tell you where my wife, Sonya, is, nor the whereabouts of my three children Sophie, Georgia and Luke. I can put odds on Sonya being in her office at work, out shopping or having a coffee with friends, but to spare my marriage I won't list those chances. Luke will likely be at college, but if the weather is nice and he doesn't have a lecture he could either be at the skatepark or out taking pictures for his photography project. Electrons live less exciting lives than members of my family, but chemists can assign them odds of being at any location around their nucleus much in the same way I can assign where my family members are likely to be.

A typical atom of hydrogen consists of one proton that forms the nucleus, and one electron. The wave function that describes where the electron could be can place it anywhere around the nucleus, but says it is much more likely to be between a range of intermediate distances, rather than being

Image of a Hydrogen Atom

close in or further out. The electron's energy level determines its wave function, its orbital around the nucleus, and consequently the distance the electron is likely to be from the proton. Electrons with more energy are found in more distant orbitals than those with less energy.

Electrons are attracted to the nucleus, and they like to be in the closest possible orbital to it. To do this, they need to have as little energy as possible. An atom of hydrogen consisting of a proton and a highly energetic electron in a distant orbital loses energy, and the electron disappears from the distant orbital and appears in a closer-in one. The energy the electron produces is carried away from the atom by a photon. As this happens, the electron vanishes from the outer orbital and appears in a more inner one.

Electrons have a property called spin, and this creates patterns in how multiple electrons arrange themselves in atoms of elements with more than one electron. Spin is all about the way an electron moves, but its definition is mathematically quite complicated. There are two types of spin – spin up and spin down – and a good analogy is to think of a spin up electron as spinning clockwise and a spin down electron as spinning anticlockwise. But like 'charm' and 'colour' in quantum theory, 'spin' doesn't have its ordinary English meaning. Electrons don't really spin like a cricket ball.

Each orbital can contain a maximum of two electrons, but two electrons with the same spin cannot share an orbital. Orbitals are arranged around atoms in structures called shells. The shell closest to the nucleus contains one orbital, the second contains four orbitals, the third nine, the fourth sixteen, the fifth twenty-five and the sixth thirty-six. In terms of numbers of electrons, shells one through six can respectively hold a maximum of two, eight, eighteen, thirty-two, fifty and seventy-two electrons.

Hydrogen has one electron in its first shell, while helium, the second-lightest element, which consists of a nucleus of two protons and two neutrons, has two electrons in its first shell – one spin up electron, and one spin down. Lithium, the next-heaviest element, has a nucleus of three protons and four neutrons, and consequently has three electrons, because each atom has the same number of protons and electrons. It contains two electrons in its first shell, and one in its second. Oganesson has two electrons in its first shell, eight in its second, eighteen in its third, thirty-two in its fourth, thirty-two in its fifth, eighteen in its sixth, and eight in its seventh. At least, that is what chemists predict. No atom of oganesson has existed for sufficiently long for its electron configuration to be accurately determined.

The lighter elements fill up their inner shells completely before moving on to the next-closest shell to the nucleus. Heavier elements fill up their shells in more complicated ways. For example, oganesson's 118 electrons are distributed across seven shells but could fit into six. Heavier elements fill up their shells in more complex ways because shells further away from the nucleus are split into subshells, and the amount of energy needed for an electron to inhabit an orbital differs between subshells. Sometimes it takes less energy for an electron to be in a lower-energy subshell in a shell further away from the nucleus than in a higher-energy one closer to it. Having written all this, I can see now why my school chemistry curriculum glossed over the details and ran with the inaccurate solar system analogy. Atoms are complicated. Yet all this complexity allows chemists to predict how atoms of different elements will react with one another, and these reactions are fundamental for our existence.

Atoms are most stable when they have as little energy as possible. An atom in this ground state has each electron inhabiting the lowest-energy orbital available. If an atom is energized such that one electron is pushed into a higher-energy orbital, it will revert to the ground state as soon as possible by releasing an electron. A hydrogen atom with its one electron in the second shell is not at the ground state, and the electron will want to lose energy by releasing a photon, enabling the electron to disappear from the outer shell and to appear in the innermost one. The photon that is released has a wavelength determined by its energy level. In the case of an electron in the second shell of a hydrogen atom, the photon released as the atom reverts to the ground state is in the ultraviolet spectrum that we cannot see with our naked eyes. Less energy is released when an electron moves from a hydrogen atom's third shell to its second shell, and the photons that are released when this move happens are consequently less energetic. They have a longer wavelength and appear to us as being red.

Even if atoms are at their ground state, they can still be reactive. Atoms are least reactive when shells, then subshells, and then orbitals are full of electrons. A lone electron in a shell makes some elements highly reactive. Hydrogen, lithium, sodium, potassium, rubidium, caesium and francium are all elements with a single electron in their outermost shell when they are in their ground states. They are found on the left-hand side of the periodic table. Hydrogen is a gas at room temperature, but all these other elements are solid, and collectively they are called the alkali metals. If an alkali metal donates the electron in its outermost shell to another atom, it becomes a positive ion. What this means is it has more protons in its nucleus than electrons in its shells and this gives it a positive charge. A positive ion is called a cation, and it will be attracted towards negative charges.

Fluorine, chlorine, bromine, iodine, astatine and tennessine are each one electron short of having a full subshell or outer shell and are known as the halogens. They are found on the right-hand side of the periodic table. Unlike the alkali metals, rather than wanting to lose an electron, they want to gain one, but when they do this they become an anion with a negative electrical charge. They are attracted to positive charges. What happens when atoms of alkali metals and halogens meet? An electron is donated from the metal atom to the halogen atom, and a compound is formed. The compound is formed via a type of chemical linkage known as an ionic bond, a strong form of electrostatic bond. Strictly speaking, compounds formed solely by ionic bonds are not molecules, although this distinction is frequently ignored in everyday English. Sodium chloride is a compound, but you will be more likely to know it as salt.

Sodium chloride can be made by mixing the elements sodium and chlorine together. Sodium atoms donate the sole electrons in their outer shells to fill up the outer shell of chlorine atoms. The sodium atom becomes positively charged, while the chlorine atom becomes negatively charged, and they are attracted to one another. When lots of ions of sodium and chlorine ionically bond, salt crystals are produced. If you make salt in the laboratory from chlorine gas and sodium metal you often need to provide a bit of energy to get the reaction started, but once it is underway and pairs of sodium and chlorine atoms transfer an electron, the reaction generates energy that can be seen as a bright yellow light. The amount of energy required to get a chemical reaction started is called its activation energy.

Ionic bonds are only one type of chemical bond. Another way of joining atoms is via something called covalent bonds, where atoms share electrons. Water molecules consist of two atoms of hydrogen and one atom of oxygen that are covalently bonded. The bonds result in each of the three atoms filling their outer shells, albeit by sharing electrons. The more electrons that are shared between atoms, the stronger the covalent bond. Carbon atoms illustrate this nicely as pairs of these atoms can bond by sharing either one, two or three electrons. It is much harder to prise apart carbon atoms when they share three electrons than it is when they share two, and it is harder to prize double-bonded carbon than single-bonded carbon atoms. When atoms of elements form covalent bonds, they produce molecules.

Chemists categorize covalent bonds into two types, polar and nonpolar. Bonds between pairs of carbon atoms are described as nonpolar when the electrons are shared equally between the two atoms. The wave function that describes where an electron is likely to be is symmetrical between the two atoms. If we could force the shared electron to appear, half the time it would be closer to one of the atoms, and the other half of the time closer to the other. Nonpolar bonds describe cases where you are on average more likely to find an electron closer to one of the atoms than the other. Water is such an example. Each shared electron is more likely to be closer to the oxygen atom than a hydrogen atom.

Nonpolar bonds mean that some parts of molecules tend to have positive or negative charges. These charged areas are called dipoles, and oppositely charged dipoles can attract, forming weak electrostatic bonds. Within water, the oxygen end tends to be a negatively charged dipole because it tends to hog the shared electrons, while the hydrogen atoms tend to be a positively charged dipole. The positive charge from the hydrogen atom allows it to form weak bonds with negative dipole oxygen, and these types of bonds are called hydrogen bonds. Although weak compared to ionic bonds, hydrogen bonds are relatively strong for an electrostatic bond. They are also extraordinarily important for life. If chemistry didn't work like this, you and I would not exist.

Hydrogen bonds are the reason that ice floats in water. When water is liquid the atoms within it are constantly moving. They have sufficient energy that hydrogen bonds do not form for long and the molecules are frequently colliding with one another. As water cools, and energy is lost from the liquid, it starts to form a lattice. This happens because the hydrogen atoms in one water molecule form hydrogen bonds with oxygen atoms in other water molecules, resulting in a twelve-sided lattice forming. The hydrogen bonds hold the water molecules apart, resulting in a solid with lots of gaps between molecules. The large amount of empty space in ice makes it less dense than water. Without hydrogen bonds ice would not float in water, or in your gin and tonic. It is relatively rare for compounds to behave like this. Most of us don't spend much time thinking about the behaviour of water, but it is surprisingly unusual.

The final property of an atom that determines how it

reacts is its size, as this influences how easy it is for an atom to share, donate or accept an electron. The more protons and electrons an alkali metal has, the more easily it will donate electrons. Potassium is more reactive than its lighter counterpart sodium because potassium can give up its sole electron in its outer shell with less effort. In contrast, the smallest halogens most readily accept an electron, and the most reactive halogen is fluorine. Because it is small, and all its electrons are close to its nucleus, it is very good at accepting electrons from other compounds, and this makes it more reactive than its cousin chlorine, which is heavier.

Fluorine is so reactive it is found in two of the most dangerous chemicals on the planet. The world's strongest acid is fluoroantimonic acid, which is a mixture of two molecules that contain fluorine: antimony pentafluoride and hydrogen fluoride. The strongest acid we are likely to come across in our everyday lives is sulphuric acid, which itself can be pretty nasty and capable of burning skin. Fluoroantimonic acid is close to 10 quadrillion times stronger than sulphuric acid (that's 1 followed by 16 zeros). It reacts violently with almost anything and is extraordinarily difficult to store, requiring containers made of compounds like Teflon that cannot be corroded.

Chlorine trifluoride is a second fluorine-based compound that is highly reactive. It is made from three atoms of fluorine and one of chlorine that are covalently bonded. Its electrons are arranged in a way that makes it highly unstable. You can use it to burn asbestos, a compound that was widely and effectively used to prevent fires in the middle of the twentieth century. It can even set alight ash made from burning wood. Although highly reactive, the compound is used in rocket fuel and in the computing industry in small doses. In everyday life, it is considered safe at levels of 0.1 parts per million, which is very dilute. There was therefore considerable risk when 900 kilograms of pure chlorine trifluoride were produced in the 1950s. While being moved across a warehouse, the chemical was accidentally spilled, burning through 30cm of concrete and 90cm of gravel before all the molecules had reacted and the fire burned out.

Highly reactive molecules are rare in nature because they quickly react with other molecules to form something more stable. Life does not use highly explosive compounds like chlorine trifluoride, 1-diazidocarbamoyl-5-azidotetrazole or fluoroantimonic acid. Although life requires energy, and highly reactive molecules can easily react, releasing it, life has found much more manageable ways of using chemistry.

If physics is the universe's way of turning energy into atoms, then chemistry is the cosmos's way of transforming elements into life. Life is picky with which elements it uses. Iron makes up just less than a third of our planet's weight, but less than 0.01 per cent of your body's. Oxygen is the second most abundant element on Earth, making up 30 per cent of its mass, but it is even more abundant in the human body, making up 65 per cent of your mass. Silicon, magnesium, sulphur and nickel are some of the most abundant elements found on Earth, yet life uses them sparsely, with each making up only a fraction of a per cent of you. We might be made from star dust, but life has been selective with which bits it uses.

Carbon is an element that is absolutely essential to life on Earth, and likely to life elsewhere in the universe if it exists. By weight, you are 18.5 per cent carbon. A carbon atom has six protons and six electrons, and anywhere between two and sixteen neutrons. The majority of carbon atoms have six neutrons and come in the isotope known as carbon-12, with only two other isotopes, carbon-13 and carbon-14, found in nature. Carbon-14 is weakly radioactive, decaying into nitrogen-14, a property that archaeologists use to age ancient artefacts. All other isotopes of carbon have only been made in laboratories and are unstable.

When carbon is in its ground state, it has two electrons in its inner shell, and four in its second shell. It can consequently make up to four covalent bonds with other atoms by sharing electrons. Being a small atom, the covalent bonds that carbon forms are quite strong, allowing it to form stable compounds. Each atom can combine in several ways to create compounds with very different properties. For example, at high pressure, each carbon atom can form single bonds with four other carbon atoms in a triangular pyramid structure, a tetrahedron, to produce diamonds. In contrast, at low pressure (and the right temperature) graphite forms. In this molecule, which is grey-black in colour, each carbon atom joins with three other carbon atoms via a complex electron-sharing arrangement. A sheet of linked carbon atoms is created, and these sheets are held together in a stack with electrostatic bonds, structured in a similar way to a ream of paper.

Carbon atoms not only join with other carbon atoms but can also form covalent bonds with atoms of other elements, including some metals. In the natural world, carbon often bonds with nitrogen, hydrogen and oxygen atoms, and because it can form multiple covalent bonds it frequently forms the large molecules that are so central to life.

There is no molecule more synonymous with life than deoxyribonucleic acid. A single molecule of DNA can contain billions of atoms. Each DNA molecule contains some of the genetic code to assemble organisms, be they bacteria, plants, fungi or animals. These molecules are stable, and DNA does not react with other molecules to change its atomic structure, which is a good thing, because if it did, the self-assembly manual to build you and me would be rapidly lost. Many other of life's molecules are similarly quite stable. We store energy in the form of molecules of fat that can be kept until they are needed to free up energy. They are stable in our cells until we need energy, and they can then be broken down. Proteins, which are the workhorses of life, are also stable, being able to repeatedly facilitate other reactions. The chemistry of life is all about controlling reactions to either break down molecules to generate energy, or to use energy to build cells and structures like bones and brains.

Like all chemical reactions, those needed for life have an activation energy. Molecules that easily react with one another form reactions with very low activation energies, while other reactions, such as splitting water molecules into their hydrogen and oxygen components, require more energy. Once a chemical reaction is underway, it can either give out energy or take it in from the environment. An energy-generating reaction releases heat and light and is termed exothermic. In contrast, one that uses heat is endothermic. If you burn something, you are creating an exothermic reaction, while photosynthesis, the chemical reactions that plants use to live and grow, is an example of an endothermic reaction because it uses photons, the energy carrier particles, to power the production of sugars that the plant needs to grow.

Life uses both exothermic and endothermic reactions, and it uses a neat trick to reduce the activation energy of many of them. It does this using a class of chemicals called enzymes. Enzymes are nearly always proteins, molecules that consist of long strings of smaller molecules called amino acids that are folded into large, complex structures. Trillions of enzymes are active throughout your body at any one time. They are responsible for joining molecules together to build your body, and also to break molecules apart when burning food to provide the energy you need to go about your daily business. Enzymes are remarkable, and without them life could not exist.

Enzymes are large and complex, with complicated three-dimensional shapes. They often have valleys and channels and odd-shaped protuberances. A small number of amino acids within these structures form something called a binding site, which is specific to each enzyme and is where the chemical reaction it facilitates occurs. The binding site allows the enzyme to form electrostatic bonds with a specific chemical that is key to the reaction the enzyme catalyses.

A catalyst is any compound that can speed up a chemical reaction, and they do this by making it easier for the chemical reaction to start by lowering its activation energy. Enzymes are the catalysts that life most frequently uses, but catalysts are also important in inorganic chemistry. The molecule that an enzyme breaks apart or builds is called the substrate, and the enzyme alters the substrate's shape. Some enzymes manage the chemical reaction alone, while others use other molecules called cofactors. The enzyme binds to the substrate, lowers the activation energy of the reaction, triggers the reaction, and then releases the end products that might be new building blocks for your body, fuel to run each of your cells, or molecules to be disposed of. The enzyme itself is not chemically altered by the reaction and survives to repeat the reaction it catalyses again and again and again.

The substrate can only bind to the active site of the protein in one way. In some enzymes, the substrate fits snuggly into the site, a bit like a 3D jigsaw. In other cases, either the substrate or the enzyme, or both, temporarily change shape a little to enable a tight bind. The binding is usually caused by hydrogen bonds, or other types of weak electrostatic bond. However, some enzymes form temporary nonpolar, and occasionally polar, covalent bonds. Once the active site and the substrate are bound, various processes can occur to start the reaction. Some of these processes involve complex chemistry that is beyond what I can cover here, but others are more easily understood.

When the substrate is forced into a different shape by the enzyme, this puts stress on the covalent electron bonds in the substrate, making them easier to break either by a cofactor, or another part of the protein that can donate or steal protons or electrons. In other cases, the enzyme orients the substrate in such a way it is easier for other molecules involved in the reaction to align and react or join with the substrate. Regardless of the process, the activation energy for the reaction is reduced and the reaction is speeded up.

The increase in the rate of the chemical reactions that enzymes catalyse can be quite remarkable, and sometimes they speed things up by several million times. This is nicely demonstrated by some work on an ancient fossil. Sixty-eight million years ago a female *Tyrannosaurus rex* died. We don't know how, but it was one of only a very small number of animals that became fossilized. Remarkably, even some of its muscles and tendons survived. In 2007, Mary Schweitzer, a molecular palaeontologist at North Carolina State University, analysed the tissue and exposed a little of it to an enzyme called collagenase. The enzyme broke down molecules of collagen in the soft tissue in a matter of hours.

Collagen is a tough protein that is central to building a body. It is found in skin, cartilage, tendons and bones. Without collagenase, it is difficult to break collagen down, which is one reason why cartilage and tendons can remain tough and stringy even after meat is cooked, and why these parts of animals are some of the slowest to be broken down by microbes following death. Despite collagen's known tough properties, Schweitzer's findings of microscopic remains of soft tissues lasting millions of years were controversial when published, and they remain so. Some independent researchers have been able to replicate some of her findings, while others have not. Science often proceeds like this. Claims that challenge current understanding need multiple independent tests before they are accepted. By no means all researchers in molecular biology or palaeontology are ready to accept Schweitzer and colleagues' findings, but my reading of the subject is that the pendulum is swinging in her favour. A growing number of scientists, including me, are becoming convinced.

I chose to describe these *T. rex* findings here, despite their being less strongly supported than the rest of the science described in this chapter. If these findings stand the test of time, they will show that an enzyme can make a reaction happen to a substrate that has been stable for nearly 70 million years. Prior to Schweitzer's findings, collagenase was known to accelerate the breakdown of collagen by many tens of times, but the *T. rex* findings suggest it can be even more potent.

Once an enzyme like collagenase has catalysed a reaction it must release the product. Any temporary chemical bonds between the enzyme and the products that have been produced must be broken. Sometimes the rearrangement of charge on the modified substrate may be enough for them to separate and disperse away, but in other cases something more is required. Following the reaction, the enzyme can return to its original, pre-binding shape, helping expel the new product. The enzyme is once again primed to conduct the reaction with a new substrate molecule.

Chemical reactions are governed by the electromagnetic force. If the electromagnetic force was substantially stronger, or weaker, electron shells and the positive and negative charges of cations and anions would be different, and life may not be possible. If atoms held on to their electrons more or less strongly, many chemical reactions would not occur or would require much higher temperatures and pressures to take place. We owe our existence to chemistry, and chemistry works as it does because of the workings of the electromagnetic force.

Although I have only scraped the surface of chemistry in this chapter, I have touched upon how atoms behave, how they share electrons, and why chemical reactions occur and molecules form. These processes are fundamental for you and me to exist. The chemistry of life is not as flamboyant as some of the reactions we encountered that involve highly reactive and explosive compounds, but it is nonetheless astonishing. Scientists have a very deep understanding of how all four forces of nature work, and their hard-won insights into the electromagnetic force means they can predict the outcome of very many reactions before they mix the compounds involved together. Nonetheless, there are aspects that are not well understood.

Chemists cannot yet easily predict many aspects of organic chemistry, the chemistry underpinning life. For example, we do not understand how many reactions that happen within living organisms are controlled and prevented from continually happening. Another thing we do not know, and this is true for all four forces of nature and not just electromagnetism, is why the forces take the strengths they do. Why is the electromagnetic force such that electron shells are spaced as they are, and not closer, or further, from atoms? Is it inevitable that the electromagnetic force is the strength it is, or is it a fluke of our universe? What we do know is that the electromagnetic force defines the chemistry we experience, and that, without chemistry, life could not exist. But it is not the only force that life required to be just right. We needed the weaker force of gravity to allow our solar system to form, and the Earth within it. I now turn to the role of gravity in our existence.

Our Locale

Our universe is truly magnificent, yet one aspect of it is deeply frustrating. It is so large that the prospects of us exploring beyond our immediate backyard are slim. The space probe *Voyager 1* was launched in September 1977 when I was nine years old, and it has been travelling for over forty-five years. It is the most distant human artefact from Earth, being about 24 billion kilometres away. A beam of light fired from Earth could reach *Voyager 1* in just less than a day. On the scale of the universe, 24 billion kilometres is very close by. Alpha Centauri, the closest star system to our sun, is 4.37 light years away. *Voyager 1* is not travelling towards Alpha Centauri, but if it was, at its current speed it would take about 80,000 years to get there.

Our sun is a star, and it, along with the planets around it, forms a star system. Our star system, the solar system, has just one star, the sun. Many star systems have more than one star, and they orbit one another in elaborate dances. The Castor star system, for example, has six stars.

Star systems, such as the solar and Castor systems, are grouped together into galaxies, with our home galaxy being the Milky Way. It contains at least 100 billion stars, and perhaps as many as 400 billion. Each star we can see twinkling in the night sky with our naked eyes is within our galaxy. It is possible to see our nearest galaxy with the naked eye on clear, moonless nights, but it is not possible to distinguish stars within it without a powerful telescope. There are estimated to be about 1 trillion galaxies in the observable universe, each separated by vast tracts of outer space. That is a lot of stars, and a lot of cosmic real estate that we can't visit or communicate with. If there are intelligent aliens out there, they would have to be cosmically close by to know of our existence.

Guglielmo Marconi sent the first radio message – 'Can you hear me?' – about 125 years ago, in 1897. If an alien civilization was listening out for radio messages and heard this and was to immediately reply, 'Yes, we can hear you,' and we were to pick up their reply, they would have to be within 62.5 light years of Earth. Radio waves travel at the speed of light, so Marconi's signal from Earth would have reached the most distant aliens who could have heard it sometime in 1959, and their response would be arriving back on Earth around now. There are about 15,000 stars within 62.5 light years of Earth, and we've heard nothing intelligent from any of these systems. Conversations with aliens on their home worlds will at their fastest take generations.

There is one radio signal astronomers have detected that may have originated from an extraterrestrial civilization. It is called the 'Wow!' signal, and it was recorded in 1977 by a radio telescope belonging to Ohio State University. The astronomer Jerry Ehman wrote 'Wow!' on the paper printout that contained the signal, and the name stuck. The signal sounds nothing like Marconi's first radio message and instead is in the form of a type of wave nothing the like of which has been detected since. In 2012, humanity responded to the Wow! message by sending 10,000 tweets towards the constellation Sagittarius, the direction from which the signal came. If the signal was from aliens, let's hope they don't think we're trolling them. Personally, I will remain doubtful that the Wow! signal is evidence of extraterrestrial life until we hear the same or a similar message again and can rule out other potential causes.

The reason I find the vast distances between star systems frustrating is I would dearly love to know how common life, and in particular intelligent life, is in our universe. The fact we have not encountered aliens does not mean they are not out there. Even if on average each galaxy contained only one planet that evolved intelligent life during its entire history, that would still equate to 1 trillion planets with civilizations in the observable universe, yet we would never know. It seems I am not destined to visit with aliens on other planets, but fortunately there is much to see on our own, and the history of our planet is amazing and has generated some breathtaking places to visit.

One of my favourite places on Earth is the Bungle Bungle mountain range of Western Australia. I became fixated on this corner of our planet as an undergraduate student. It was the late 1980s, and my friends and I were avid watchers of the Australian soap opera *Neighbours*. One character, Helen Daniels, announced occasional trips to the Bungle Bungles to paint, presumably whenever the actress who played her, Anne Haddy, needed a break. I didn't believe there could possibly be a place called the Bungle Bungles so headed off to the library to learn more. After spending way more time than I had expected, and time I could ill afford, given my looming exams, I learned that the Bungle Bungles was an ancient mountain range consisting of beehive-shaped striped mounds and canyons carved by wind, water and sand over millions of years. The area had been first seen by Europeans in 1983 when a film crew flew over it, and four years later it was designated a National Park. The local Gija people had been visiting Purnululu, their name for the range and what it now should be referred to as, for over 20,000 years, and possibly much longer.

It took me thirty years to make the trek to Purnululu, with my wife Sonya, and my children Sophie, Georgia and Luke, visiting for my fiftieth birthday in 2018. To get there, we flew on progressively smaller planes from London to Singapore to Darwin to Kununurra before the final leg on an eight-seater propeller plane which deposited us on a sandy airstrip that is grandly named Purnululu airport, close to the range. Although it took me three decades to visit this UNESCO World Heritage-listed site, I was told by someone who had a dream job of working in the park that I had fallen for an Australian marketing campaign. Having gazetted the national park in 1987, the Australian Tourism Board wanted to attract visitors to it, and although I have been unable to independently verify this, embedding the Bungle Bungles into Helen Daniels' lines was apparently part of the campaign.

The beehive-shaped mounds of Purnululu are striped orange and black, and they sit in front of mountains that today reach a height of only 575 metres above the surrounding plains. These mountains are criss-crossed by canyons carved over the course of millions of years. The history of the mountain range dates to a time when amphibians were the dominant animals, well before mammals had evolved. Between 350 and 375 million years ago, multiple layers of sediment were laid down in the bed of a river or lake before they were eventually compressed into layers of sandstone. Geological activity caused by the shifting of huge slabs of rock called tectonic plates that form the crust of our planet then pushed the sandstone upwards to form a towering mountain range. Over time, the range was eroded by rainfall, streams, wind and fluctuations in temperature, and Purnululu was formed. The orange and black stripes of the mounds have faced the brunt of winds blowing off what is now the Simpson Desert, and they are the most eroded part of the ancient range. The stripes are thought to be due to differences in the clay content of the sedimentary sandstone layers from which the range initially formed. The more clay that was in the sediment when it formed, the higher the water content of the rock, and these wetter layers are home to cyanobacteria, a species of photosynthesizing microorganism. The stripes are only a couple of millimetres thick because the cyanobacteria need light to survive and it does not penetrate far into the rock. The bacteria are also unable to survive in drier layers of the rock. The stripes are consequently indicators of fluctuations between wet and dry periods that took place over a 25-million-year period nearly 400 million years ago in the ancient lake where the sediment was laid down.

The Purnululu range is unique to Earth, and Earth itself is unique. In the topics we have covered so far in this book, we have not yet had to focus on local events. Gravity, electromagnetism and the weak and strong nuclear forces operate throughout the universe in the same way (or at least physicists assume they do because they emerged when the universe was very tiny and are thought not to have changed, and there is no evidence to suggest they vary from one galaxy to the next). Matter is not spread evenly throughout the universe due to quantum variation when the universe was very young, and that difference in its distribution resulted in gravity pulling together matter into stars and galaxies. The four fundamental forces are responsible for our planet and the solar system in which we reside, but to understand how our little corner of the universe came to be we must now think a little more locally. The details on how matter has been distributed in our neck of the woods become important as we consider how our solar system and planet formed. Was it inevitable that a star the size of our sun developed where it is today, or that it would have planets tracing the orbits we observe across our night sky? Each star is a little bit different, as is each solar system and galaxy. For you and me to exist, we needed the right combination of events to occur locally so that a planet formed where life could evolve and develop into an intelligent species. In this chapter, I largely focus on what we know about how our local conditions came about, and gravity is the force that takes centre stage. But before I get on to that, there is a gap to fill.

The fundamental forces, the first stars, heavier elements and the first molecules were all in existence before the universe was a few hundred million years old. The universe formed 13.77 billion years ago, yet our solar system formed only 4.6 billion years ago. What happened in the intervening billions of years?

Galaxies formed about 13.6 billion years ago, and they exist in groups and clusters across the heavens. The Milky Way is one of at least fifty-four galaxies in a cluster called the Local Group, which in turn is one of a hundred or so groups that form the Virgo supercluster of galaxies, which itself is a neighbourhood of the Laniakea supercluster of 100,000 galaxies. The smallest of these galaxies contain only a thousand stars or so, while the largest contain 100 trillion or more.

Galaxies move. All galaxies spin around a central point because of the way they formed from vast clouds of hydrogen gas being pulled together by gravity. On average, galaxies are also moving apart as the universe expands. However, galaxies within clusters like the Local Group can move towards one another, and they can collide. Astronomers have discovered 30,000 or so stars in the Milky Way with orbits that go in the opposite direction to that of most of its stars, and they have identified these as being part of another galaxy which crashed into the Milky Way many eons ago. Calculations of the trajectories of the Milky Way and the Andromeda Galaxy, a close neighbour only 2.5 million light years away, suggest they will collide in 4.5 billion years' time.

Although all galaxies began to form at about the same time, they don't all die at the same age. The Hubble Space Telescope, a remarkable device orbiting Earth that has provided many insights into our universe, has identified six galaxies that died by the time the universe was 3 billion years old. For reasons unknown, they had used up all their hydrogen and could no longer form new stars. In contrast, most galaxies are still producing stars, although the rate at which they do varies. Some galaxies contain lots of heavy elements like iron, having burned through many generations of stars, while others contain few heavy elements, being on star generation one or two. Within the Milky Way, some areas are more metal and oxygen rich than others, revealing that the history of different parts within a galaxy can also vary.

Most of the first stars in the Milky Way have died, but many new stars continue to be born. Parts of our galaxy consist of dust and molecules from the death of past stars, along with vast amounts of hydrogen and helium created in the early universe but not yet captured by stars. Sometimes these vast clouds are shaken into action by the explosion of stars several times more massive than our sun. Such supernovae are some of the most violent events in our universe, and they can result in dust and gas clouds forming nebulae, the birthplaces of new stars. The Helix Nebula is Earth's closest star-forming nursery, with light from it taking seven centuries to reach us. In nebulae such as these, gravity causes cold gas and dust to start to clump together, becoming warmer and eventually forming new stars. These processes occur throughout the galaxy where there is sufficient hydrogen, with small, medium and large stars being produced, many with planets, but a few without.

Averaging across the star systems studied to date, each star has between one and five planets. Each of the star systems and each of its planets is unique. Although their formation has been driven by gravity, variation in the density and composition of the dust in the nebula in which they were born, and the strength of the star's gravitational pull at the location where each planet formed, determine the planet's size and composition. Each star and each planet has its own history, so now to turn to the history of the solar system.

Between 4.9 and 5 billion years ago a supernova triggered the formation of our solar system. If the nebula in which our sun formed was like those we observe elsewhere in the galaxy, many other stars and solar systems formed within it. Once young stars and solar systems form in nebulae, the spin of the Milky Way sets them off on their first orbit around the centre of our galaxy. At the centre of the Milky Way is an enormous black hole weighing about 4 million times the mass of our sun. We call the black hole Sagittarius A*, and the stars and solar systems of our galaxy orbit this black hole, at least until they are sucked into it, or they burn out, having used up all their hydrogen fuel. Although large, there are some black holes that are an order of magnitude larger than Sagittarius A*: in early 2023 astronomers estimated that a black hole called TON618 is ten times larger.

Our solar system is 25,640 light years from the centre of

the galaxy. Two hundred and fifty-six centuries ago, humans built the first settlement, consisting of huts built of rocks and mammoth bones. Light that left the sun at the time of these innovative ancestors of ours will only now be arriving at the centre of the Milky Way. Given how far it is to the centre of the galaxy, it is not surprising that it takes a long time for our solar system to make a complete orbit of the galaxy. So long that scientists got fed up with dealing with such large numbers and instead defined a new measure of time to describe the orbit: a galactic year. Each galactic year is between 220 and 230 million years (it's hard to measure), and the Earth is about twenty galactic years old. The first life evolved about 16.9 galactic years ago, while the first humans made their appearance approximately a galactic fortnight ago.

The Sun's Position in the Milky Way

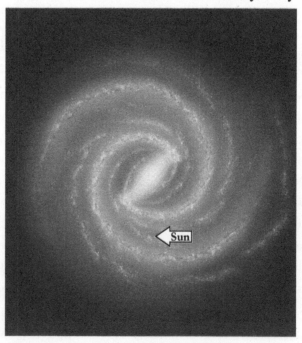

By 4.6 billion years, or 20.4 galactic years ago, our sun had formed, with huge amounts of hydrogen being pulled together by gravity until nuclear fusion began. The sun is by far the biggest in the solar system and constitutes 99 per cent of its mass. Although tiny compared to some of the objects I've discussed in this chapter such as massive black holes, it is, to us, huge. It has a diameter that is eleven times larger than Jupiter's, the next-largest object in the solar system, and 109 times larger than that of Earth's.

Other stars that formed in the same nebula as the sun set off on different orbits around the galaxy, with some moving faster and others more slowly. The closest stars to us did not form in the same nebula as Earth, and we know this because they vary significantly in age. For example, Barnard's Star, the second-closest star system to Earth, is about twice as old as our sun, while the Luhman 16 system, the next-closest system, is only 600–800 million years old. Because different star systems orbit the galaxy at different rates, Alpha Centauri, the closest star system to us today, will likely not be our closest neighbour a billion years hence.

Much as the trajectory of galaxies within galaxy clusters reveals each galaxy pursuing its own journey across the universe, each star system within our galaxy follows its own orbit around Sagittarius A*. Some of these orbits mean that some stars repeatedly pass close by one another. Every galactic year or so, our sun will fly by star systems it has encountered before. Some scientists have argued that such close encounters may result in meteoroids and asteroids being displaced from their orbits in the outer reaches of our solar system, putting them on new trajectories that take them closer to the sun, potentially putting them on a collision course with Earth. There is evidence that asteroid and meteorite impacts peak and trough on galactic year timescales, although not all scientists are persuaded. Collecting more data to confirm this happens will take a very long time.

The terminology of space rocks can be confusing. I have mentioned asteroids, meteoroids, meteors and meteorites, but how do they differ? An asteroid is a rocky object that has a diameter of greater than a metre and that is smaller than a planet. A meteoroid is also rocky but is less than a metre in diameter. When a meteoroid enters the Earth's atmosphere it burns up, producing a meteor that appears as a shooting star across the night sky. If the meteor hits the ground it is classified as a meteorite. As an asteroid approaches a planet, it often breaks up into meteoroids, which become visible in the night sky as a meteor storm. The impact that killed the dinosaurs was an asteroid that didn't disintegrate completely into meteoroids, and which resulted in a large lump of rock crashing into the planet. In contrast to an asteroid, meteoroid, meteor, meteorite or space rock, a comet is made of ice and dust rather than rock. As comets in our solar system near the sun, they partially melt, leaving a trail of water and dust. Light reflecting off the melting comet makes it appear as a bright light travelling across the heavens.

We do not have enough data from other star systems to know whether our solar system is typical or atypical. Perhaps solar systems such as ours are two a penny, or perhaps ours is the consequence of a set of highly specific circumstances that have never been repeated. The study of exoplanets – planets orbiting other stars – reveals that planets are common, and planets of a similar size to Earth that orbit around stars at a distance where water, a key chemical for life, exists as liquid, are likely not particularly rare. Many

Earth-like planets may exist, but that does not mean we are not unique. Earth's history is important on the journey from the Big Bang to you and me, and to study it we have to piece together our planet's history using all the tools of the scientific method. At this point, I take a slight detour from the history of our planet to explain the next bit of my history of becoming a scientist.

I found the science taught at school was largely uninspiring, and things didn't immediately improve when I headed to university in the delightful Cathedral city of York in the north of England. I loved university, but many of the lectures left me bored, and I continued to largely self-learn in the library. I enjoyed reading papers published in scientific journals. In one lecture in the second year of my biology course, a friend and I were sitting at the back being disruptive. We were chatting a bit too loudly and flicking balls of paper at our friends. At the end of the lecture we were asked to remain behind so the professor, John Lawton, could tell us off.

I felt mortified, so later that day I went to Professor Lawton's office to apologize. He asked me why I had been fooling around, and I explained that I didn't feel challenged by the lectures and that the contents of the scientific papers I was reading seemed much more nuanced. Professor Lawton told me I was right but that we had to be taught the basics, which was fair enough, and that I should hang in there as I would find the research projects of our final year much more stimulating. He proved right, but I never saw him at York again, and I quickly forgot about my telling-off.

A year later, with the end of my undergraduate course approaching, I needed to decide what to do next, and, other than understanding why I existed, I really had no idea what

I wanted to do. My then girlfriend was applying to do a Ph.D. at Imperial College London, so I decided to apply too. I didn't have a clear idea of what graduate studies might entail, but I put a lot of effort into my application and sent it off. A few weeks later I was invited to interview, and the letter informed me that I would be interviewed by Professors Mick Crawley and John Lawton. My heart dropped. Professor Lawton would recognize me, remember my disruptive behaviour, and that would be game over for my hope of graduate studies at Imperial. Except it didn't play out like that. John, as I came to know him, greeted me with a 'hello again, Tim' before explaining there was no need to mention our past interaction. The interview went well and I was offered a place. John had remembered me from our conversation and had been impressed with my comments about the lecture content and my extracurricular reading. I wouldn't recommend mucking around in class, but on that occasion it paid dividends.

John and Mick felt I should also be advised by Professors Charlie Canham and Steve Pacala for part of my Ph.D. that involved data collection from a large project the two Americans ran in the north-east of the United States. I didn't realize it at the time, but John, Mick, Charlie and Steve constituted a dream team of scientific mentors, not only for their stellar track records but also because they each took a different approach to their science. John is an all-rounder: an exceptional naturalist with a deep understanding of how the natural world works. He would frequently tell me to 'take a step back and think about the big picture' when I got caught up in the minutiae of my own project. I doubt he ever imagined I would take quite so many steps back to write this book. Mick studies plants and is an expert in how to use statistical approaches to analyse biological data to test hypotheses. He is also a talented teacher, being able to enthuse the innumerate about statistics. Charlie is a forest ecologist who studies nutrients and energy flow through food webs, and Steve is a theoretician who builds and analyses models to acquire general insight.

A little bit of each of their approaches rubbed off on me, and I developed into a science all-rounder rather than a champion of any single element of the scientific method. Their mentorship also taught me that good scientists are imaginative but can discard promising ideas that violate existing understanding, can alter their opinions when data shows a cherished hypothesis is wrong, and are evidenced-based and constructively critical. Science has been said to advance one funeral at a time, because many scientists will not give up strongly held beliefs even in the face of overwhelming evidence, but my mentors showed me that didn't need to be the case. Mick used to reject a hypothesis annually. After finishing my Ph.D. I worked with him and other collaborators on a remote Scottish island that was home to a population of wild sheep. Each year we would have a bet on the number of sheep alive on the island before counting them, and I developed a predictive mathematical model that worked reasonably well. Predicting the future was not Mick's greatest strength, but he would cheerily accept his hypothesis was wrong and grudgingly pay up. He did well to choose science as a career, as he'd not have excelled as a Wall Street trader.

When studying the local history of Earth, or even a small part of it, it is necessary to use the full toolbox of the scientific method. By gazing into space, we can look back in time and see what was happening in galaxies billions of years ago, but we cannot train our telescope on the ancient Earth.

Geologists and astronomers have pieced together the history of our planet through the cunning application of all aspects of the scientific method, and it is a field of study that I loved researching for this book. There have been many disagreements among geologists as they pieced together the history of our planet, from how the old Earth might be to whether the continents move or not. These arguments led to new data being collected, and we now have a very good history of our planet. With this short detour over, I return to the history of Earth.

The planet on which we live, with its continents, oceans and diverse forms of life, has not always been thus. At various times in its past, the Earth has had a surface of molten rock, has been covered in a thick layer of ice, has been much hotter than it is today, and much colder. It has been battered by meteoroids, asteroids and space rocks and has even collided with a planet that was about the size of Mars. Our lives are too short to experience more than a tiny fraction of a geological epoch, and apart from the occasional earthquake and volcanic eruption, the geology of our planet appears stable and calm to us, which can be a source of comfort given the turbulence of human existence. But geological stability on the timescale of our lives is an illusion. Our planet is dynamic and is always changing, but the rate of change is slow compared to the three score years and ten of the average human life. Yet each year my hometown of Oxford moves about an inch further away from one of the places where I work, Yellowstone National Park, as the Atlantic Ocean grows wider due to plate tectonics, the slow movement of the great slabs of rock that form the surface of the Earth. Similarly, the tall mountains of Papua New Guinea, where I wish I worked, get taller with each passing year, while the Pacific Ocean shrinks.

4.6 billion years ago, hydrogen and other elements along with ice and other molecules in the nebula that was the birth-place of our sun began to collapse under the force of gravity to form a large disc with the young sun in its centre. The disc likely contained rings, a little like those seen around Saturn today. Dust and ice in these rings began to join, or in cosmic parlance accrete, into larger objects, with these pebble-sized meteoroids combining and colliding until small-planet-sized rocks developed. The collisions and the actions of gravity increased the size and temperature of these rocks and they combined into a few larger small planets called planetesimals orbiting the sun. These planetesimals steadily accreted more material from the disc, becoming planets. Eventually, at a distance of 150 million kilometres from the sun, a molten rocky mass formed that was the baby Earth. By 4.543 billion years ago the Earth existed. Back then, it was not prime real estate, and any form of life would have quickly met a grizzly end.

Similar processes happened at different distances from the sun. Our solar system has four rocky planets – Mercury, Venus, Earth and Mars – that are closest to our star. Further out, there are the four gas giants Jupiter, Saturn, Uranus and Neptune, and these are all beyond the frost line, a distance from the sun beyond which water forms ice and gaseous compounds condense into solids, creating the gas giants, while within it only heavier compounds can be accreted, forming rocky planets. Planets that formed close to the sun are rocky because they have a greater proportion of heavier elements than those that formed further out. It is a key reason why each of the planets that form the solar system are so different from one another.

The early Earth was about 40 million years old when it collided with another planet named Theia. The collision heated the slowly cooling Earth, creating a surface of molten rock, and it would have thrown a large amount of dust and debris into the atmosphere and beyond. Some of this debris fell back to Earth, but the remainder accreted to form the moon. The young moon orbited the Earth at a much closer distance than it does today, probably at between 25,000 and 30,000 kilometres. Tidal forces for the first oceans when they formed must have been colossal with the moon close by, yet with each orbit it has slowly moved further away, and today it is currently 384,400 kilometres distant.

Collisions between celestial objects were more common between 3.9 and 4.5 billion years ago than they are now. Many of these collisions were between the young planets and the myriad meteors that littered the solar system. Collisions with meteorites were particularly important for Earth, as this is where most of the water in our oceans came from. Although geologists have found evidence of past meteorite and space-rock impacts on Earth, including the site of the impact that ushered in the end of the Cretaceous geological epoch and the extinction of the dinosaurs, much of the evidence of the frequency of past impacts comes from studying the surface of the moon and Mars. Scars from past meteorite and space-rock impacts disappear on Earth at a much faster rate than they do on the moon and Mars. The reasons for this are that the moon and Mars do not have plate tectonics caused by shifting slabs of rock, surface water that can erode rocks, or plant roots and bacteria that over millennia can crumble and break rocks apart. The scars on these other celestial objects reveal impacts between planets, moons, asteroids and meteoroids are frequent on astronomical timescales. This history is hard to read on Earth because it is steadily eroded over time.

Jupiter, the fifth and largest planet in the solar system, formed at about the same time as Earth. Computer simulations of the youthful solar system suggest that shortly after it formed it began a journey towards the sun, only ceasing it at about the distance that Mars is now from our star. It stopped its journey towards a fiery collision course with the sun because Saturn had developed an orbit that resulted in a complicated dance between the two large gas giants, resulting in Jupiter and Saturn establishing orbits close to where they are now found. Jupiter is about three and a half times further away from the sun than the Earth is, while Saturn is even further distant. This account of early planetary meanderings, which took millions of years to complete, is called the Grand Tack hypothesis.

As Jupiter Grand Tacked towards the sun and back, its gravity prevented another planet forming. The asteroid belt, which lies between Mars's and Jupiter's current positions, would have accreted into another planet had Jupiter not trekked towards the sun and back. Jupiter's journey to the inner solar system also resulted in dust, gas and debris either being pushed further out into the solar system or colliding with the gas giant itself, resulting in Mars being significantly smaller than it would have been otherwise. Without Jupiter's wanderings, Mars would be larger than Earth, yet it is much smaller. In some star systems, astronomers have observed large, rocky planets that orbit so close to their star they are at risk of being consumed by it. It is possible that Jupiter's Grand Tack prevented Earth realizing this fate, by stopping it from accreting more material and moving closer to the sun.

We tend to think of our solar system as being fairly static in its behaviour, and certainly it is on the span of a human life, or even over a few million years. On longer timescales our solar system is dynamic and the behaviour of one planet can influence the others. Without Jupiter's wanderings early in its life our planet might not have settled to an orbit in the habitable zone – the range of distances from a star where water exists as a liquid and not as steam or ice. Without young Jupiter's exotic orbit, it is possible that Earth would not have remained in the habitable zone and life would not have emerged.

The Earth's orbit around the sun takes 365.24 days, an Earth year. The planets closer to the sun take less time to make one of their orbits, while those further away take longer. Neptune, the most distant planet, takes nearly 165 Earth years to orbit the sun. Pluto, which was a planet when I was a child but got downgraded to a dwarf planet when scientists learned it had failed to accrete other objects in its immediate neighbourhood, takes even longer: 248 years pass while it makes one orbit of the sun. Not only do the planets furthest from the sun have the greatest distance to travel to complete an orbit, they also travel more slowly. Mercury moves through space at a little under 110,000 mph, while Neptune moves nine times more slowly. Earth, by comparison, hurtles through space at 66,615 mph, or 0.006 the speed of light.

The distance from the sun to the Earth, nearly 93 million miles, is defined by scientists as one astronomical unit. Mercury, the closest planet, orbits the sun at 0.387 astronomical units, while Neptune is just over thirty astronomical units from our star. Life exists on Earth, but is that because it is just the right distance from the sun, or could there be life on any of the other planets or their moons? Astronomers have long been interested in this question, and they initially addressed it through estimating the range of astronomical units at which water would exist as a liquid.

Early estimates suggested that the habitable zone was narrow, and Earth was rather lucky to sit within it. More recently the width of the habitable zone has been extended, as the probes we have launched to explore our solar system have revealed evidence that water may exist as a liquid on both Europa, a moon of Jupiter, and Enceladus, a moon of Saturn that is 9.5 times further away from the sun than Earth. These findings complicate estimates of the size of the habitable zone around a star because it depends not only on how far a celestial object is from the star, but also on the atmosphere and the internal geology of the object, as well as the gravitational pull of other celestial objects. Mars has long been thought to be in the habitable zone, but it has a very thin atmosphere and liquid water is not found on its surface. In contrast, Europa is covered in a layer of ice 10–15 miles thick. Under this, liquid water almost certainly exists in an ocean 40–100 miles deep. The liquid water doesn't freeze, due to the complex gravitational pull of Jupiter and two of its other moons, Io and Ganymede. All three of these moons have eccentric orbits around their mother planet. These orbits result in a tidal pull that heats the oceans so they do not freeze. Europa may also experience volcanic geothermal activity, injecting further heat into its oceans. The compelling evidence for the existence of liquid water on Europa and Enceladus has led to scientists speculating that these may be the most promising homes for life beyond Earth in our solar system. Who knows, perhaps alien fish swim in these oceans?

Most planets and moons within the habitable zone do not support life. There is no life on the moon, and even if life did once exist on Mars, it is unlikely it has persisted to the modern day. All life on Earth requires liquid water and it is widely assumed that extraterrestrials, be they single-celled bacteria or multicellular little green men, will also require it. At sea level on Earth, water freezes at zero degrees Celsius and boils at 100 degrees, but at higher altitudes these temperatures change. On top of Mount Everest water boils at only 68 degrees Celsius, making it hard to have a decent cup of tea, while at the bottom of the deepest ocean trenches the water pressure is so great that water boils at 400 degrees Celsius. As pressure decreases, the freezing point of water also declines. To understand the range at which water exists as a liquid in a particular location it is necessary to know the pressure there, and on the surface of a planet this is determined by the composition of its atmosphere.

The atmosphere of a planet is most dense at its surface, and it thins with altitude. The Earth's atmosphere is divided into five zones. At the surface, and to an average height of about 7.5 miles, is the troposphere, the layer in which we live. It is where the vast majority – 99 per cent – of our atmosphere's water vapour is found. The stratosphere comes next, extending to an altitude of 31 miles. The stratosphere contains the ozone layer, which protects us from much of the solar radiation that can cause sunburn and replication errors in the genetic code that can lead to cancers. Ozone is a molecule consisting of three atoms of oxygen, and when humanity's pollution created a hole in the ozone layer in the

1990s we realized how our impact on the planet permeates way beyond the surface where we reside. After the stratosphere comes the mesosphere, which extends to 50 miles above sea level, and this is where most of the meteoroids heading for the Earth's surface burn up as meteors. Temperatures drop to −85 degrees Celsius in the higher reaches of the mesosphere. The next layer, where the International Space Station is to be found, is called the thermosphere. It extends to about 450 miles above the Earth. Molecules are starting to get very sparse up here. The final level is the exosphere, and it extends to over 6,300 miles above us, with the molecules found here being at risk of being lost to outer space. Most of our satellites orbit the Earth in the exosphere. After the exosphere it is another 378,000 miles to the moon.

Today, the troposphere consists of 78 per cent nitrogen, 21 per cent oxygen and 0.04 per cent carbon dioxide, with the rest made up largely of water vapour and argon. Over the history of the Earth, the composition of the troposphere has changed. During the age of dinosaurs, carbon dioxide was five times more common than it is now, oxygen was at 27 per cent, nitrogen at 70 per cent, and the average humidity would have been about twice what it is today. If dinosaurs had sweat glands (they didn't), they would have been very sweaty. Going back even further, the atmospheric layers we see today would not have existed. The very early molten Earth would have had no atmosphere at all, but after the collision with Theia and as it aged and cooled an atmosphere of methane, hydrogen sulfide, carbon monoxide and carbon dioxide formed, with the latter making up as much as 10 per cent of the early air. As the Earth continued to cool and experienced frequent meteorite impacts, the content of water in the atmosphere increased, until the first pools and oceans began to form about 4 billion years ago.

Atmospheric pressure on the surface of our planet has also varied with time. Scientists have used models to predict that 4 billion years ago the Earth's atmosphere would have been thick and heavy, similar to that of Venus today. It then started a trend to reduce in pressure over time, but with fluctuations. Some estimates suggest the atmosphere on the surface of the Earth was four times as dense as it is now on the day the dinosaur-destroying asteroid hit, while examination of tiny bubbles of gas in 2.7-billion-year-old larva suggests that the atmosphere back then may have had only half the pressure we experience today. There is a lot of uncertainty around these estimates, so treat them with caution. Studying the early atmosphere is not straightforward, but geologists agree that Earth's temperature and atmospheric pressure had created conditions where liquid water could exist by about 4 billion years ago.

Although both the atmosphere and distance from the sun play a major role in determining whether water exists in liquid form, other factors are also at play. On Earth, night-time temperatures are lower than those during the day, as the dark side of the Earth cools when not bathed in sunlight. The same is true of some other planets. On Mercury, daytime temperatures peak at over 450 degrees Celsius, while night-time temperatures plunge to -170 degrees Celsius. Part of the reason for the huge range of temperatures is the length of the Mercurian day. Each day on the closest planet to the sun is 176 Earth days. A night-time of nearly three months is sufficient for a large drop in temperature. Venus, too, has long days, with one Venusian day lasting 243 Earth days. However, the temperature on its surface remains constant day and night at 465 degrees Celsius. The difference between the temperature fluctuations on Mercury and Venus is due to the composition of their atmospheres. As the concentrations of greenhouse gases such as carbon dioxide and methane increase in an atmosphere it becomes harder for heat to escape. Venus has more greenhouse gases than Mercury. Earth has a relatively low concentration of greenhouse gases in its atmosphere compared to Venus, although humans are changing these concentrations. If Earth spun on its axis more slowly, such that it had days that lasted for months, day and night temperatures would reach extremes that could have made it impossible for complex life to evolve.

Temperatures not only fluctuate on daily timescales but also on annual, and even longer ones. Much of the Earth experiences summer and winter as a function of how each location on Earth faces the sun. Both the North and South Pole experience weeks of darkness because the rest of our planet comes between them and our star. This is because Earth spins on a tilted axis in relation to the sun, and it is this tilt that causes seasons in temperate and polar zones. The way the moon orbits the Earth means that we always see the same side of it. If you were to live on the dark side of the moon, you would never get to see our green and blue planet. If the Earth's axis were tilted such that parts of our planet never saw the sun, it is highly probable that some parts of it would be unsuitable for life.

The tilt of the axis of a planet can change over time. Earth's axial tilt, or obliquity as it is formally called, oscillates between 22.1 and 24.5 degrees on a 41,000-year timescale. In contrast, Mars's obliquity ranges from 15 to 35 degrees. Even the relatively small scale of the Earth's obliquity can cause climate cycles. Extreme obliquity could have made conditions challenging for life, particularly complex multicellular life, to arise.

In the last couple of million years, Earth has experienced climate troughs and peaks on a 100,000-year cycle. The obliquity contributes to these cycles, but so too does change to something called the eccentricity of Earth's orbit. When we think about the Earth orbiting the sun we typically think of it tracing a perfect circle, but that is not right. All orbits are elliptical, but some ellipses are closer to circles than others, and it is fortunate for us that Earth's elliptical orbit is reasonably circular. Its departure from a circular orbit is described as its eccentricity. As the distance from the sun increases during this orbit, less sunshine reaches the Earth and temperatures can drop, while during times the orbit takes the planet close to the sun, things warm up. This long-term fluctuation in temperature should not be confused with the annual winter/summer fluctuation we feel at high latitudes. The eccentricity of Earth's orbit changes with time, and this is a major driver of the 100,000-year climate cycles that have driven ice ages on Earth over the last 2.5 million years. If variation in our planet's eccentricity were greater, it could be possible that Earth would periodically leave the habitable zone, negatively impacting life on Earth.

Despite all the things that could have prevented it, our planet lies in the habitable zone around our sun, not too far away and not too close. It has a stable orbit that is not too eccentric, and it spins on an axis and at a rate that keeps water liquid across much of the globe. Even small changes to these parameters might have meant oceans, rivers and lakes did not form, and instead the Earth might have been a ball of ice, or a hothouse where the only water was in the atmosphere as steam. Yet it is not just the atmosphere above our heads and our position in the solar system that has helped life thrive but also the solid land beneath our feet.

The Earth is approximately spherical, and consists of layers, a little like a scotch egg. If you have never eaten a scotch egg, you should. It consists of a boiled egg surrounded by a thick layer of sausage meat which in turn is covered in breadcrumbs, before the whole delicious sphere is deep-fried. Like a scotch egg, our planet has four layers. The centre, analogous to the yolk of the egg, consists of a solid inner core made primarily of iron. Surrounding this is a molten outer core, which consists of iron and other heavy metals. The egg white represents the outer core. The outer core generates the Earth's magnetic field, another key feature of our planet. Surrounding the outer core is the mantle, analogous to the sausage meat layer of the scotch egg. The mantle is a rocky layer consisting of silicates, molecules containing some atoms of silicon and oxygen. On top of the mantle sits the crust, the thinnest layer – the breadcrumbs on the outside of the scotch egg. The breadcrumbs give a scotch egg a rough texture, and you might think this parallels rugged mountains on Earth's surface. Intriguingly, however, if you could shrink our planet to the size of a scotch egg, it would feel as smooth as a billiard ball. Even tall mountain ranges like the Himalayas and the Andes would not interrupt the smoothness on a shrunken-down model of the Earth the size of scotch egg.

The Earth's magnetic field acts as a shield that helps protect our atmosphere and has been essential for life. Without the magnetic outer core, life may have long since died out on Earth. To understand why, we need to revisit the sun. It is not only planets that have atmospheres, so too do stars, and

Anatomy of the Earth

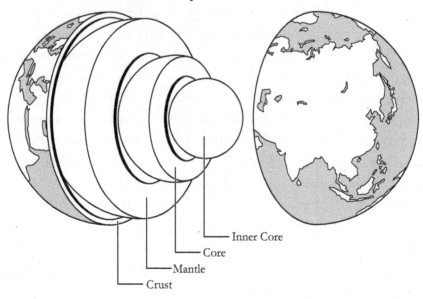

Inner Core

Core

Mantle

Crust

our sun is no exception. The outermost layer of the sun's atmosphere is called the corona, and for reasons that we only partially understand it is extraordinarily hot, much hotter than the sun's surface, reaching a temperature of a million degrees Celsius. One reason it may be so hot is because of the sun's magnetic fields. At such a high temperature, electrons are ripped away from their atomic nuclei and a plasma of charged ions is formed. Particles in this plasma are then accelerated to very high speeds by the sun's magnetic field, before they form solar winds that hurtle out through the solar system at speeds of between 200 and 500 miles per second.

When these charged ions collide with gas molecules in a planet's atmosphere, the collisions can result in the gas molecules being jettisoned into outer space. Mars loses a greater proportion of its atmosphere to solar wind than Earth. Over the last 4 billion years, the solar wind has stripped Mars of the thicker atmosphere it had once had. The atmosphere on Mars is now thin as there is little of it left.

For the first half a billion years of Mars's life the solar wind did not ravage the planet's atmosphere so destructively. Back then, Mars had a magnetic field, a bit like Earth has today. If a planet has a magnetic field, it can deflect the solar wind, protecting the atmosphere from such significant losses. On Earth, the solar wind enters only the upper two layers of our atmosphere, the thermosphere and exosphere, but when it does it can cause spectacular sights: the Northern and Southern Lights. These remarkable light displays, caused by collisions between the sparse gases in the higher atmosphere with the protons, electrons and helium nuclei that form the solar wind, are a result of our planet's magnetic field reducing the rate at which gas is lost from our atmosphere to outer space. The obvious question is why does the Earth have a magnetic field, and why did Mars have one which petered out so early in its existence?

The outer core is the egg-white layer of Earth in my scotch-egg analogy. It consists of molten iron, silicates, sulphides (molecules containing atoms of sulphur) and an unknown quantity of radioactive metals. Humans have not visited the centre of Earth, and there is much we do not know about it, but the circulation of molten iron in the outer core creates what scientists call a dynamo, and it is this that is responsible for our planet's magnetic field. If the outer core were to solidify or shrink, the magnetic field that protects our atmosphere would shut down and our atmosphere would slowly be lost to outer space. This raises the question, why is the outer core molten?

Some of the core's heat is left over from the formation of

the Earth. The accretion of material from the early solar system and the collision with Theia generated lots of heat. As the outer layers of our planet cooled and solidified as rock, the core became insulated, and heat only dissipates slowly. The action of gravity and the decay of radioactive metals will also generate heat (although the relative contribution of each process is not known). There are consequently multiple things happening that have kept our outer core molten, and as long as it remains thus the Earth will have a magnetic field. Mars's outer core, being smaller than the Earth's, cooled and solidified earlier in its history. Its dynamo lasted only about 500 million years. From that point on, its atmosphere was destined to be stripped by the solar winds, and life, if it had evolved, was not to thrive as it has on Earth.

The molten core means the Earth is a huge magnet and, just like a smaller magnet you might attach to your fridge, it has a north and south pole. The magnetic north pole, where northern lines of attraction of the Earth's magnetic field enter the planet, is currently on Ellesmere Island in Canada. The magnetic south pole is just off Antarctica, in the direction of Australia. These poles slowly meander, and 600 million years ago may have been close to the equator, rather than being nearer the poles. Many times in the history of Earth the magnetic field has even switched orientation, with magnetic north becoming magnetic south, and vice versa. The history of these switches can be detected in various rock deposits around the globe. Scientists have been unable to identify any pattern in these switches. They can occur as frequently as once every 10,000 years, or as infrequently as every 50 million years, but on average occur every 300,000 years. The last switch in polarity occurred 773,000 years ago, and may have taken 7,000 years to complete. The field has become close to flipping a further fifteen times since, but it hasn't quite happened. There were several species of bipedal apes alive when the last reversal happened, including the first members of the *Homo* genus (our relatives) living in Europe. I suspect they were oblivious and largely unaffected by our planet's changing magnetic field. We would notice such a change now as it would play havoc with our electronic devices. As well as switching poles, the Earth's magnetic field also fluctuates in strength with time, sometimes being stronger and other times weaker.

Above the molten outer core sits the mantle – the sausage meat of my scotch-egg planet. The mantle consists primarily of rock-forming minerals made up of magnesium oxide and silicates. These rock-forming minerals have names like olivine, garnet and pyroxene. Although the mantle is solid, on timescales of millions of years it moves ever so slowly. Nearly 85 per cent of the Earth's volume is made up of the mantle. It is at least 1,750 miles thick and is hot, but not as hot as the core of our planet. The deepest part of the mantle, next to the outer core, reaches a temperature of 4,000–5,000 degrees Celsius, while the top of the mantle is considerably cooler, at 200–600 degrees Celsius.

The temperature gradient between the top and bottom of the mantle generates movement where the hot minerals and rocks deeper down very slowly rise towards the surface while the cooler material closer to the surface moves inwards. It can take hundreds of millions of years for a piece of material to move from the top to the bottom of the mantle. A human lifespan is a blink of an eye compared to the eons over which the dynamics of the Earth's mantle play out. Despite this, the slow convection is important for our planet's dynamics and drives plate tectonics that can lead to earthquakes and some types of volcanic eruptions. These are the most dramatic home-grown events our planet experiences, but before discussing them it will be helpful to consider the outermost layer of the Earth, the breadcrumbs covering the scotch egg: the crust.

The Earth's crust is fractured and consists of fifteen tectonic plates. These are large slabs of rock, and they are divided into seven major and eight minor plates. The crust ranges from being 3 miles thick under the oceans to an average of 20 miles thick on continents, although in some places where there are large mountains such as the Himalayas, it can be as thick as 60 miles. Our neighbouring planet Mars is not as fractured as Earth and it has no plate tectonic activity at all. Venus is about the same size as Earth, and up until recently scientists thought that its crust was much thicker and that it too was devoid of plate tectonics. However, using recent observations from the *Magellan* spacecraft, researchers have revealed that Venus's crust is of a similar average thickness to Earth's. The scotch-egg structure of our planet may be typical of rocky planets, and that means Earth may eventually share the same fate as Mars. In time, as our planet cools, the crust will harden and the movement of tectonic plates will cease. But for the foreseeable future our fractured crust will continue to be dynamic, continents will shift and earthquakes will be a risk in many parts of the world.

New crust is continually being formed in areas where two tectonic plates move apart on deep-sea ridges. As the plates move apart, molten rock from the mantle, called magma, moves up, before cooling and forming new crust. In areas where two plates are being pushed together, one plate is pushed down, slowly melting, before joining the mantle and starting its long journey towards the core. The tectonic plate that is pushed up creates land. This is why the continental crust where land occurs is thicker than the plate under the sea.

Continental crust gradually gets worn away via processes that geologists call weathering and erosion. Repeated freezing and thawing of water in fissures in rocks can break them apart, wind and rain can wear them down over eons, and acids in water can also contribute to weathering. Life, too, can contribute to weathering and erosion, with plant roots, fungi, bacteria and even some animal species contributing to the breakdown of rocks. Over long time periods erosion can reduce the heights of tall mountain ranges. Think of the Purnululu range that I visited for my fiftieth birthday.

Tectonic plates are continually shifting and changing size and shape depending upon whether the rate of loss to the mantle or the rate of gain from the formation of new crust is faster. If our crust was thicker, convection in the mantle would fail to generate the plate tectonics that fracture plates or drive them apart, and it is possible there would be no, or very little, land on our planet. One consequence of our planet's thin crust is that the pressure on tectonic plates can build up as they rub up against one another, and this can lead to earthquakes as the pressure gets too much and the tectonic plates suddenly jerk. The ground shakes as the plates shift and settle into new positions. Tectonic activity has been a key feature of the history of land on our planet, and played an important role in the spread and diversification of life that led to our existence. For example, 200–300 million years ago there was only one supercontinent, Pangea. Plate tectonics broke it apart and, slowly, the continents that are familiar to us emerged. The break-up of Pangea split apart populations of animals and plants, and in many cases as time ticked by these populations diverged, evolving to become completely different species on separate continents. In more modern times, the movement of tectonic plates can be catastrophic for humans. The largest earthquake in the twenty-first century occurred in the Indian Ocean on 26 December 2004. The resulting tsunami killed over 225,000 people.

Fifty-four million years ago a tectonic plate that contained what is now Europe and North America split in two. The reason for the split is unclear, but it may have been the weight of the two continents being pulled towards the centre of the planet by gravity, along with the build-up of magma at a weak point in the centre of the plate. When the ancient plate split it caused enormous volcanic activity, throwing vast quantities of hot rock, ash and dioxides of carbon and sulphur into the atmosphere. Such fractures of a plate are, mercifully, rare, but they do not always generate massive volcanic activity. Sixty-three million years ago, another plate split into the ones where modern-day Seychelles and India are found. Volcanic activity in the region 6 million years earlier meant that little magma had built up and was insufficient to cause a huge eruption like that seen between what is now modern-day North America and Europe 9 million years later. Comparisons such as these reveal that local history matters. Each earthquake, tectonic plate fracture and volcanic eruption is different. Yet they are driven by the same processes, such as the movement of rocks in the mantle, new crust formation in the deep sea, hardening of the crust, and subduction, where one plate is forced beneath another as they collide. These processes interact, generating a range of possible geological outcomes. Earth scientists have developed a remarkable understanding of these processes and are starting to gain astonishing insights from other planets, but accurately predicting what will happen in the future is currently hard: we cannot predict earthquakes or volcanic eruptions in advance with much confidence.

At the time when the first life evolved, Earth's atmosphere was made from the gases ejected from volcanic eruptions. Hydrogen sulfide, methane, carbon monoxide and carbon dioxide were abundant. If life had not evolved, our atmosphere's composition today might resemble that found on Venus, consisting of 95 per cent carbon dioxide and only a percentage or two of nitrogen, oxygen and other molecules. Not that we'd be around to measure it. Although the exact composition of the atmosphere early in Earth's history is not known precisely, scientists agree that the percentages of carbon dioxide and methane have dropped significantly over time, while nitrogen and oxygen have increased. Oxygen didn't start to increase until between 2 and 2.5 billion years ago. It was produced as a by-product of photosynthesis in species of cyanobacteria, but it could not accumulate in the atmosphere in any appreciable amount until other reactive elements, such as iron, had oxidized. Nitrogen started to increase in abundance from about 4.4 billion years ago. It is so abundant in the atmosphere today because it does not easily combine with other elements to form the crystals and rocks that make the ground beneath our feet. Nitrogen is consequently common in the atmosphere and relatively rare in the planet. Oxygen, in contrast, is found in rocks because it oxidizes with many other elements so easily, and it also forms water.

Atmospheric scientists predict that in about a billion years' time oxygen levels in the Earth's atmosphere will drop catastrophically, leading to the extinction of all multicellular life. The reason for this is that carbon dioxide levels will drop to such low levels, partly because our sun will have got hotter, and this will result in photosynthesis (the predominant source of atmospheric oxygen) no longer being possible.

As the Earth's atmosphere has changed, so too has its average global temperature. There have been five spells when the Earth has been cold, so-called Icehouse Earth epochs. These chills lasted millions of years, repeatedly causing major extinctions. Yet some of these icehouse spells also led to periods of evolutionary innovation. The Sturtian glaciation lasted for nearly 60 million years in a period known as snowball Earth, ending about 660 million years ago. It may well have been a catalyst for the evolution and subsequent spread of multicellular life. During some icehouse Earth periods glaciers advanced and retreated, but areas of ice on the planet are a consistent feature of our planet when it is in this state. We are currently in an icehouse earth epoch (there is ice at the poles and on mountains), but we are between ice ages, meaning there is not as much ice as there was 15,000 years ago or likely will be 50,000 years hence. Whether this ice age happens or the planet leaves the current icehouse spell may well depend on humanity's production of carbon dioxide and other greenhouse gases in the coming decades and centuries. Exactly what triggers major icehouse Earth events is unclear, but it could be variations in the Earth's orbit, changes in the composition of the atmosphere, or even changes in the amount of volcanic activity.

Our planet has also experienced periods of sweltering heat, with the Earth being in its greenhouse state for about 85 per cent of its existence. Today, the average global temperature is 13.9 degrees Celsius, but during greenhouse periods the average temperature has hit 30 degrees. When you hear these numbers, you may wonder what all the fuss about anthropogenic climate change is about. Life will undoubtedly survive even a rapid change in climate of a few degrees, but many species will not. Humans might survive, but we would doubtless suffer, as our civilizations are fragile and susceptible to collapse.

In much of the past, the climate changed relatively slowly, over hundreds of thousands or millions of years. But occasionally there would be much more rapid change caused by asteroid impacts or major volcanic eruptions. The Siberian Traps are evidence of a period of massive volcanic activity that lasted for 2 million years, about 250 million years ago. Temperatures can change quite quickly when volcanoes start erupting.

The word 'trap' is a geological term used to describe a type of rock formation. Nearly 3 million square miles of modern-day Siberia, an area about the size of Australia, is covered with basaltic rock that was formed from ancient volcanic eruptions. Huge amounts of lava were ejected from a massive build-up of magma under the Earth's surface. Not only did these eruptions cover a vast tract of land with molten rock, they would also have spewed huge quantities of carbon dioxide and other greenhouse gases into the atmosphere, along with ash and other debris. The ash and debris in the atmosphere would have temporarily reduced sunlight, lowering the temperature on Earth. Once the ash and debris settled, the carbon dioxide would have remained, being accumulated in plants, sediments and the ocean relatively slowly, acting as a greenhouse gas that warmed the planet. Estimates suggest that the Siberian Traps produced between 12,000 and 100,000 gigatons of carbon dioxide, where a gigaton is a billion tonnes. The carbon dioxide that was released increased the concentration of carbon dioxide in the atmosphere by four times, which was enough to raise the temperature of the

Permian atmosphere by 10–15 degrees Celsius and the temperature of tropical oceans to 40 degrees Celsius.

The Siberian Traps continued to erupt for approximately 2 million years, although most of the activity may have occurred in about a quarter of that time. These eruptions resulted in four out of five ocean-living species being driven extinct, and 70 per cent of terrestrial species being lost forever. Tetrapods, the animal group to which humans belong, were very nearly driven extinct and are thought to have survived on only a small part of the ancient continent Pangea.

Volcanic eruptions can cause home-grown mass extinctions, but the cause of the best known of the mass extinctions came from the heavens. If you gaze at the night sky on a cloudless evening in the absence of artificial light, you can see shooting stars. The best places to see these are in areas such as deserts during colder nights, where there is little haze in the atmosphere. The first time I saw the night sky on a clear, cold night from my father-in-law's farm in rural Queensland was remarkable. The Milky Way traversed the sky, with countless stars. Through binoculars, I could see another galaxy. Meteors in the higher reaches of our atmosphere caused shooting stars every few minutes, and I was transfixed. If I had grown up in a place like this, I might well have become a space scientist, but by the time Sonya and I got together I was already a professor of zoology.

Most of the rocks on a collision course with Earth are small, and they burn up in our atmosphere as meteors I watch at the farm. Larger ones can make it further into our atmosphere. On 30 June 1908, a 100,000-kilogram rock travelling at over 60,000 mph entered Earth's atmosphere, causing what Earth scientists refer to as the Tunguska event. Much of the 50–60-metre-wide rock burnt up as it came through the atmosphere, but at a height of 5 miles above the Earth it shattered into small pieces, causing a meteor airburst. As the meteor collided with our thicker atmosphere it caused a large explosion, which in the case of the Tunguska event flattened 80 million trees in Siberia over an area of 2,150 square kilometres. In this case, humanity escaped significant tragedy. Had the meteor impacted the Earth's lower atmosphere over a built-up area, millions could have died.

Some meteoroids do not burn up but make it all the way to the planet's surface. Their fate depends upon their size, their angle of approach to Earth and their composition. In October 1992, a meteorite hit Earth in Peekskill, New York, destroying a car. It partly disintegrated during its transit and the core that hit the surface of the Earth weighed 890 grams. It had been travelling through our solar system for 4.4 billion years, ending its journey by hitting the boot of a twelve-year-old Chevy Malibu. It really is the journey and not the destination that matters. I was unable to find out whether the seventeen-year-old owner of the car, Michelle Knapp, was able to claim against her insurance, but I suspect not, given the behaviour of the insurance companies I have had dealings with over much more earthly accidents. I do not feel too sorry for Michelle though: Wikipedia reports that she sold the meteorite for $50,000, and the damaged car, which had cost her $300, for $25,000. I imagine she has been dining out on the story ever since. I certainly would.

Our galaxy likely contains hundreds of billions of planets orbiting around its 100 billion or more stars. We have no reason to suspect that our galaxy is unique in being home to planets, with the first tentative evidence of planets orbiting stars in other galaxies even starting to emerge. If we assume that other galaxies have planets, there will be hundreds of trillions of planets in the observable universe. The four fundamental forces operate throughout space as they do in our neck of the cosmic woods, and these forces interacted to produce Earth. Each planet will have its own unique history, determined by the local density of matter when it formed, its success at accreting additional matter as it grew, its star, and its distance from the star, but given the vast number of planets in the universe it would be a surprise if there were not similar planets to Earth. These planets will be slightly different, having days and years that are a little bit shorter or longer, slightly weaker or slightly stronger masses and gravitational pulls, more or less eccentric orbits and different axial tilts. They will have experienced different climates and different meteorite impacts, but they will have oceans of liquid water, volcanic eruptions, one or more moons and earthquakes. We have yet to discover such a planet (although NASA has a list of possibilities) from 5,000 exoplanets identified to date. Five thousand is a tiny, tiny, tiny fraction of the planets in the Milky Way, let alone the observable universe.

Earth and space scientists have pieced together the history of our planet through a mix of observation, experiment and mathematical modelling, while drawing on insights from chemistry, physics and biology. They study the goings-on on other planets and star systems via telescopes and through analysis of meteorites that have collided with our planet or meteoroids that have been harvested from space by probes, and by looking at the behaviour of atoms to date rocks. Much of the history of our planet has been learned by characterizing the chemical composition of rock layers of different types of different ages and working out what conditions must have been like when these rocks were formed. There is still much to learn, but one thing we do know is that conditions at one or more locations of the young Earth about 4 billion years ago were such that life emerged. Chemistry got seriously complicated, and that is the next part of our history to consider.

The Emergence of Life

The focus of the book until now has been on the physical, chemical and earth sciences and summarizes what happened, and what had to happen, to create a planet where life could emerge. The next few chapters move on to focus on the life sciences, on how life arose and diversified, and how some species became very complicated. DNA is a chemical that is central to this story. Its first remarkable property is that it can make copies of itself. A second, even more remarkable property of DNA is its ability to encode instructions to make complex objects such as proteins, cells and bodies that aid its replication. Some of these objects are staggeringly complicated, and some of them help DNA codes replicate better than others. Evolution is the natural process that selects the DNA codes that are most effective at replicating themselves. What had to happen for you to exist is DNA had to come into existence, and evolution had to select DNA codes that result in you and me. To understand why this happened we need to know not only how life emerged, but also how the genetic code works, and how evolution invented things like multicellularity, sex, sentience and cooperative behaviour. The next few chapters consider this part of the journey from the Big Bang to us.

The first step on our journey is to explore why life emerged, and it is one of science's most challenging problems. Although the problem is difficult, it can be quite easily articulated: how did life self-assemble the first self-assembly manual that describes how to build an organism? All living organisms contain an instruction manual – a genome – that describes how to go about making itself. How did the first genome that provides the instructions on how to build an organism arise? It was not designed, but instead it came about via the interaction of molecules somewhere on the Earth when it was young. All organisms alive today are descended from this first primitive organism.

All life uses a chemical called deoxyribonucleic acid, or DNA for short. The complete collection of an individual's DNA is referred to as its genome, and every living organism on Earth, from the tiniest bacteria to the largest blue whale, has a genome. Each genome carries instructions to develop an organism. Your genome codes for a set of steps that enabled you to develop from a single cell formed by the fusion of an egg from your mother and a sperm from your father into a thinking, talking, eating, book-reading, intelligent ape. Similarly, my dog Woofler's genome codes for a set of steps to develop from a single fertilized egg an animal which many people describe as cute despite his keen desire to chase down squirrels and his irresistible urge to roll in anything that smells terrible. These self-assembly manuals are not written in words but in genes: stretches of DNA that carry instructions to make molecules called proteins that are fundamental to life. The purpose of any life, including yours (from a biological perspective at least), is to produce offspring, because those offspring carry your genes, and these genes help your offspring develop, hopefully into biologically successful adults. Life is all about making copies of itself. Every genome in every organism contains instructions to make proteins that are necessary for the genome to make copies of itself. But which came first? The genome or the proteins?

Biologists are dealing with the 'which came first' problem, the chicken or the egg, but at the level of molecules.

The emergence of life from non-living chemicals is called abiogenesis. Research into the mechanisms underpinning abiogenesis is highly technical, involving very complicated chemical reactions. Scientists studying abiogenesis have worked out how and where many of the key molecules central to life formed, but they have not yet been able to explain how these key molecules self-assembled into the first, simple, singled-celled organism. Part of the reason for this is that even the simplest single-celled organism alive today is staggeringly complicated, containing over 500 genes producing hundreds of different proteins. These simple organisms are hugely more complicated than our most complex machines, including the LHC at CERN. I am not going to describe the chemical mechanisms that emergence-of-life researchers have uncovered, but I will instead focus on what had to happen for life to emerge.

Life got going relatively quickly after our planet formed, suggesting that when conditions are right, it can evolve without too much difficulty. The oldest non-controversial traces of life are found in the rocks of the Pilbara in North-western Australia, and date back approximately 3.4 billion years. These traces are in the form of fossilized stromatolites, layers of sediment formed by simple microorganisms called cyanobacteria. Stromatolites are still around today, having possibly changed very little over the last 3.4 billion years, and they can be seen at Shark Bay, just a few hundred kilometres from the Pilbara. I visited them in 2022 – it was a bit of a pilgrimage for me – and they look like shrunken-down versions of the beehive-shaped mounds of Purnululu I described earlier, in that they are often dome-shaped and striped. However, they usually contain limestone rather than the sandstone from which Purnululu are formed, and they only grow to about half a metre tall. These days stromatolites are a rarity – the only other place they are found is the Bahamas – but in the past they were more widespread. The cyanobacteria that created ancient stromatolites were descended from simpler life forms, with the earliest ghosts of life thought to have appeared sometime in the preceding 500 million years. Rocks from Quebec and Greenland have tantalizing evidence of this earlier life in the form of specific chemical signatures, although scientists do not agree on the origins of these chemicals. Exactly when the first life appeared remains unclear, but it probably evolved when the Earth was just a few hundred million years old. You, me, the vegetables in your salad, the worms in your garden and Woofler are all descended from early forms of life, and that is why understanding how life arose is central to the story of how and why we came to exist.

The species I have just listed – even the salad ingredients – are all extraordinarily complicated and built of vast numbers of cells. Yet even the simplest single-celled form of life alive today is very complex. The simplest bacterium did not simply and suddenly spring into existence. It evolved from something simpler. These simpler forms of life are long extinct, and fossil evidence of them is lacking, so knowing what they were poses a scientific challenge. The divide between the mix of organic compounds that were the precursors of life and even the simplest bacteria is so vast that understanding the journey life has made is daunting. Scientists are yet to see life emerge out of beakers of chemicals in the laboratory, and this means there is inevitably some speculation about how life began, and how the earliest forms behaved. Nonetheless, early life would have been subject to the forces of evolution that govern it today, and so some inference as to the processes that had to happen for life to evolve can be drawn from contemporary observations.

Distinguishing between the mix of organic molecules that led to life and the first living thing is not straightforward. There is a continuum from the set of chemicals from which life emerged to you and me, and deciding where to divide this continuum into living and non-living is subjective. Scientists have long argued over where the life–nonlife break in this continuum should be because any discussion of life requires a definition. Although not really romantic enough for something as beautiful as life, I will define it as a set of naturally arising, replicating, controlled chemical reactions within a membrane that is maintained by an external energy source. Try putting that into a song lyric.

It is worth discussing this definition a little. 'Naturally arising' means that life arose from simpler chemical reactions and was not designed, produced or kick-started by some higher being. 'Replication' refers to the production of descendants. All living organisms strive to make copies of themselves via clonal or sexual reproduction. Clonal offspring are genetically identical to their parent, while those produced by sexual reproduction share half their DNA with each of their two parents. Biological and evolutionary success is measured in the number of offspring you leave behind, with each carrying replicate copies of some parental DNA. From a historical perspective Pope Innocent III and Genghis Khan were both influential leaders during their time, yet their biological legacies are very different. Unlike some popes, such as Sergius III and Alexander VI, who fathered children during their tenures as pope, Innocent III had no children or descendants, while one in 200 men alive today is estimated to be descended from Genghis Khan. He will also have many female descendants, but the way our genetics works (and more on that later) means that they are harder to count. No one knows how many children Genghis Khan had, with only nine formally recognized, but some estimates suggest as many as a thousand.

Returning to life's definition, the next part states 'a controlled chemical reaction'. This is a reaction that can be speeded up, slowed down or stopped. At any point in time, there are millions and millions of chemical reactions going on inside your body. If you are healthy and tumour-free, each is controlled and operates within bounds. Evolution has found ways of slowing or stopping reactions when the compounds produced are sufficiently abundant for living organisms to run smoothly, while initiating and speeding up others when other molecules are scarcer than life would like. These chemical reactions are powered by 'external energy' in the form of light, reactive chemicals or food, and the majority of these reactions occur either inside or on the outer 'membrane' of cells that form the basic building blocks of all forms of life.

Each cell, regardless of whether it belongs to a bacterium or a multicellular organism like you, is complicated. Countless chemical processes and reactions occur continually within each cell, including the reading of the genetic code and the production of proteins and other large and complex molecules. Even in the simplest free-living bacterium these chemical processes are the product of hundreds of genes. There are some simpler parasitic species of bacteria that have fewer genes, but only because they rely on the hosts they invade to do some of the hard work of living for them.

In this chapter, and the next, I consider how life began and spread to conquer the planet, but before discussing life's origins it is worth considering just how complicated some key aspects of it are, as that will help frame the enormous challenge scientists face when considering how even the simplest forms of life alive today came about.

DNA is fundamental to life. A molecule of DNA is a long strand constructed from four smaller molecular building blocks collectively referred to as nucleotides. A nucleotide can be further broken down into a nucleobase, a phosphate molecule and a sugar molecule. The phosphate molecule is made up of three atoms of oxygen and one of phosphorus, while the sugar is called deoxyribose and is built from carbon, hydrogen and oxygen. The four nucleobases in DNA are called adenine, guanine, cytosine and thymine, and within each nucleotide one of these nucleobases binds to the sugar. Each nucleobase consists of a few atoms of carbon, nitrogen, hydrogen and (except for adenine) oxygen. Although a large and complicated molecule, DNA is consequently made from common atoms combined in a way to form a genetic sequence. Each molecule of DNA is unreactive and stable. It is important not because it gets involved in chemical reactions (it doesn't) but rather because it contains the genetic code to self-assemble an organism. It is DNA's structure that is fundamental to life.

When scientists talk of genetic sequences, they are referring to the order of the nucleobases in a strand of DNA. In a single strand there is no restriction on how the nucleobases are ordered. Biologists usually refer to each nucleobase by its first letter such that a genetic sequence might be AGCGTG-GCCAGTC, and so on, and on, and on. However, when two strands come together to form DNA's famous double helix structure, cytosine and guanine always form bonds with one another between the strands, as do adenine and thymine. DNA's double helix is created by hydrogen bonds forming between these pairs of nucleobases. If the spiral were to be removed from this helix, DNA would look like a very long ladder. The vertical sides of the ladder are the deoxyribose and phosphates, while the horizontal slats of the ladder are the hydrogen-bonded nucleobases. Twist the ladder and you produce a double helix.

Some parts of a strand of DNA (genetic sequences) contain instructions to make proteins that are fundamental

DNA Structure

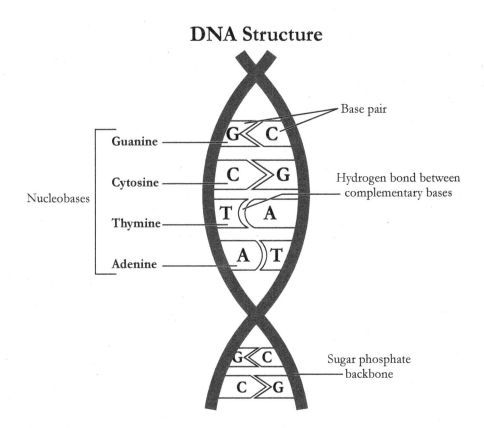

to life. Each DNA instruction to make a protein is a gene. Your genome consists of 3 billion nucleobases, with many (but not all) of these packaged into 20,000 or so genes. Your genome provided the instructions to self-assemble you, written in code to make proteins that determine how you were to develop from a fertilized egg into you, and how you should function in day-to-day life.

Proteins are three-dimensional molecules that are made by folding up chains of smaller molecules called amino acids. A gene provides the ordered list of amino acids that need to be joined together in a chain to produce the protein. Life uses twenty amino acids to build proteins, although many more types exist. A chain is started by joining two amino acids together and is lengthened by adding new amino acids to one end of the chain. The genetic code describes which amino acids are to be added at each step. Each of the twenty types of amino acid is coded by three-nucleobase sequences. For example, an amino acid called histidine is coded by the triplet of nucleobases adenine–thymine–guanine, or ATG for short, while the amino acid leucine can be coded for with eight different nucleobase combinations: TTT, TTC, TTA, TTG, CTT, CTC, CTA and CTG.

A bit like DNA, proteins are stable, and most are not particularly reactive. As an example, recall the *T. rex* protein called collagen I described that survived for 66 million years before being broken down by collagenase. Some proteins are enzymes and act to catalyse reactions, making them happen more easily and speeding them up. Collagenase is an enzyme, and it works by binding to a collagen molecule, making it easier to break it apart. In doing this, collagenase is not chemically altered. It is the structure of collagenase that allows it to do this. The structure helps collagen molecules get involved in a chemical reaction, which then results in them being broken down into smaller molecules.

The proteins that genes produce are often assembled to create more complex molecules. For example, in humans there are forty-three genes involved in making collagen molecules. I glossed over some of the details in my previous discussion of collagen, but you have twenty-eight different types of collagens. They are all involved in helping build your body, and they are built from the proteins that the forty-three genes produce. The genes produce simpler proteins that are fused to make more complex collagen molecules. Instructions to produce the collagen are coded for by genes written in DNA, but there is no single gene for the protein. Your genome, and the way you are built, is more complicated than one gene equals one protein.

Not all of your DNA codes for a gene. Some parts act as switches, turning genes on or off, and this allows an organism to regulate how much of a particular type of protein is produced. When a gene is read, or expressed as biologists say, the gene produces the string of amino acids required to make a protein. The switches tell the cellular machinery when a particular gene should be on or off. In every cell, all the time, some genes are switched on while others are switched off, and over time the list of genes that are making amino acid chains, and those that are not, changes. Feedback loops ensure that if one particular protein becomes too common within a cell, it triggers instructions to turn off the gene that produces it.

Excusing mature red blood cells, which are unusual among human cells in that they do not contain a cellular structure called a nucleus where DNA is stored, each of your cells contains the same DNA, but they do not use it in the same way. What differs between cells is not the genes they contain but the genes that are expressed. Your 220 or so cell types differ from each other because different genes were turned on and off at different times during their development. Contrasting sets of genes are also used to allow each cell to do its job. The set of genes that is turned on in your skin cells is different from the set that is turned on in your brain cells.

Some of your genetic switches will be different from mine. We will also have different genetic sequences at some of our genes, and this means that these genes are likely to produce proteins that do their jobs at different rates in you and me. These genetic differences mean that although we are both human, we are not identical. Differences in our DNA might mean you have a different eye or hair colour to mine, are taller, have different allergies, and are more or less susceptible to various diseases. On average, 99.9 per cent of my genome sequence will be identical to yours, with only 0.1 per cent differing. If I compare my DNA with Woofler's, I would find an overlap in sequences of about 85 per cent.

Your genome is unique to you, with no one else, living or dead, sharing your exact genetic sequence. Unless, that is, you have an identical sibling. The uniqueness of your genetic sequence contributes to making you different from everyone else, and these differences have arisen in part due to genetic mutations – errors in DNA replication. Some genetic mutations help the individuals that carry them survive and reproduce a little better than those individuals that lack them. In contrast, others have little impact on how individuals develop and function, while some cause genetic conditions such as dwarfism or ocular albinism. The latter is a genetic complaint from which I suffer, and it means that some cells in the back of my eyes fail to produce a coloured pigment. I know that doesn't sound very important, but it means I can see only about 25 per cent as well as someone with perfect vision. The mutation came from my mother's side of the family, and the genetic condition is much more likely to impact men compared to women.

Half of your DNA comes from your mother, and half from your father. It is tempting to think that one of the two strands in your double helix came from each parent, but that is not what happens. Your DNA is arranged into twenty-three chromosome pairs, giving a total of forty-six strands of double-helical DNA, with each strand called a chromosome. Scientists have imaginatively named these: chromosome 1, chromosome 2, chromosome 3, etc. You have two copies of chromosome 1, and two of chromosome 2, and so on, with one copy in each chromosome pair coming from your mother and the other from your father. Ignoring for now chromosome 23, the Y chromosome, which determines whether you are male, the version of each chromosome you get from each parent is unique, being a mix of each of their chromosome pairs.

When your mother produced the egg that was to become you, and your father the sperm that was to fertilize that egg, meiosis – the biological process that makes these reproductive cells – produced new versions of chromosomes 1, 2, 3, 4, etc. never before seen by life (continue to ignore the Y chromosome). For example, the unique version of chromosome 1 in your mother's egg was produced by cutting sequences from each chromosome in her chromosome 1 pair and joining them together to produce a new version of chromosome 1. In exactly the same way, the version of chromosome 1 in your father's successful sperm was produced by blending bits of your father's chromosome 1 pair.

The same thing happens to each chromosome, except for the Y chromosome as you have probably now realized, whenever meiosis produces an egg or sperm. Imagine the two versions of your chromosome 12 as blue and red threads. Line the two threads up next to one another and then cut through both with a pair of scissors. Join the top red half to the bottom blue half to make a new thread, with the top half red and the bottom half blue. That is how new chromosomes are made. In this example, a new version of chromosome 12 is produced that will be incorporated in one of your sperm or egg cells. Here, I've described cutting through the chromosomes once, but I could have cut through both multiple times. Each cut-and-join is called a chromosomal crossover, or recombination event. On average, there are between one and four crossovers per chromosome when a reproductive cell is produced, although some chromosomes experience higher rates of recombination than others.

The Y chromosome is different, and it is why it is easier to estimate the number of male descendants of Genghis Khan than the females who count him as an ancestor. The Y chromosome experiences recombination only very rarely. Chromosome pair 23 contains the sex chromosomes, and these determine whether you are male or female. In mammals the sex chromosomes are called X and Y. If you are female, you have two X chromosomes, while if you are male, you have one X and one Y chromosome. A female's chromosome 23 pair consequently consists of XX, while a male chromosome 23 pair is XY. X and Y are of very different lengths – Y is tiny compared to X – and an X and Y chromosome cannot readily recombine to make a mix of X and Y.

The X chromosome you receive from your mother is a mixture of your mother's two X chromosomes. However, your father only has one X chromosome, and if you are a female this was passed on to you un-recombined. It is identical to your father's single X chromosome. If you are male, you don't get an X chromosome from your father, instead you receive his Y chromosome. This too has not been mixed with another chromosome. No individual has two Y chromosomes, and this means no Y chromosome ever gets mixed with another Y chromosome. However, very, very occasionally a Y chromosome may mix with a bit of DNA from another chromosome, such as an X chromosome or even another chromosome in a completely different chromosome pair, but when this happens it is an error in replication. On some of these rare occasions when it occurs, it can signal the beginning of an evolutionary journey that results in a new species, but that is a topic for another book.

The above paragraphs reveal that the self-assembly manual that is your genetic code is complex. We will discover why it is so complex later in the book when I discuss why sex evolved. For now, it is sufficient to appreciate that DNA is a large molecule, and the way it is inherited, passed from parents to offspring, with your genetic sequence being a mix of those of your parents, is complicated. Upon your eventual death, you will cease to be, but if you have children some of your genes will live on. Genes replicate, being passed from parents to offspring, but individuals die.

Biologists distinguish the reproductive cells that can fuse with one another to produce the first cell of a new individual in the next generation from the non-reproductive cells that have the sole purpose of helping to keep individuals alive long enough for them to reproduce. The reproductive cells make up the germ line, while the non-reproductive cells are collectively called the disposable soma. What is truly remarkable about life on Earth is how complicated the disposable soma has evolved to be in higher animals like you and me. Evolution has created genetic sequences to self-assemble organisms with arms, legs, eyes, brains, skin, teeth, toenails, tonsils and sentience. These attributes maximize the chance that the DNA sequences in germ lines that instruct how these body parts are produced are passed on to the next generation. DNA is a truly remarkable molecule. It replicates, making copies of itself, but it does this by providing self-assembly manuals to build species as diverse as cats, cabbages, cockroaches and chlamydia.

Although not as complicated as some species, the way that single-celled bacteria, like chlamydia, make a copy of their DNA still requires many steps. There are nine enzymes that are essential for bacterial DNA replication, and many others that play less crucial roles. Skipping over the details, most bacteria species store their double helix DNA molecule in a loop. Replication takes one loop and produces two, and it happens just before a bacterium splits to become two bacteria. The process begins with the double helix being separated by enzymes at a single location on the loop. Other enzymes steadily unzip the double helix, with the point at which the strands are separated called the fork. Yet more proteins latch on to each of the separated strands. On one strand, enzymes follow close behind the moving fork, reading the nucleobases and producing a continuous mirror-image strand of DNA. On the opposite strand things are more complicated. The enzyme that moves along DNA making a mirror-image copy can only move in one direction, and in this case, it is away from the moving fork. On the second, or lagging, strand, the enzymes join close to the fork and move away from it, producing small strings of nucleotides called Okazaki fragments that then need to be stitched together by yet more proteins to produce a continuous strand of DNA. Even DNA replication in simple species, and I use the word 'simple' with a touch of irony, is very involved and did not simply spring into existence.

Replication is not the only complicated bit in life's definition. Making proteins from the DNA code is significantly more involved than I have space to explain, and it relies on another key chemical of life called ribonucleic acid, or RNA as it is abbreviated. I won't say much about RNA, but I will introduce it in a little detail as it may have played a key role in the emergence of life. Only single-stranded, with a backbone of alternating ribose and phosphate molecules, and using a nucleobase called uracil instead of thymine, it is a close cousin of DNA. Ribose has one more oxygen atom than deoxyribose, but is otherwise very similar, while uracil is simpler than thymine, and in particular lacks a group of carbon and hydrogen atoms that form the hydrogen bonds that keep DNA's two strands bound together.

Like DNA replication, and the way proteins are synthesized from the genetic code, metabolism is also fiendishly complicated, and it did not arise in a single step. Human metabolism works by taking energy in the form of sugars such as glucose from our food and using them to create a proton gradient, by concentrating more protons on one side of a molecular membrane than on the other. Protons are one of the building blocks of atomic nuclei, and were described in an earlier chapter. The proton gradient is then used to make a chemical called adenine triphosphate, or ATP, from a molecule of adenine diphosphate, or ADP, and a molecule of phosphate. ATP is important because it is the fuel that life uses to run. In animals, like me and you, the proton gradient used to make ATP is inside structures found inside all your cells called mitochondria. These organelles, the word that scientists use to describe structures within cells, are surrounded by a double membrane, and the proton gradient is created by moving protons into and out of the space between these two membranes – the intermembrane space. Mitochondria, and their inter-membrane space, are critical for life because that is where ATP is produced.

If left to their own devices, and assuming they could cross the membrane unhindered, the protons would settle to an equilibrium where the same concentration of protons would exist on either side of the membrane. But that does not happen. The protons are instead being pushed from a lower concentration inside the mitochondrion to a higher concentration in the intermembrane space. To do this, electrons are passed along a chain of chemicals called, appropriately enough, an electron transfer chain, and as they move along molecules in the chain, they change those molecules' electric charge, and this pushes protons across the membrane. Life uses the electromagnetic force to run. By the end of the process, two electrons have pushed ten protons across the membrane through special channels before the electrons are used, along with a molecule of oxygen (O_2) and four other protons, to make two molecules of water. You'll eventually pee or sweat these out of your body.

The proton gradient is then used by the cell to create ATP. The positively charged protons in the intermembrane space readily attach to a negatively charged part of a protein called ATP synthase that sits in the mitochondrion's inner membrane. The protons then pass through a channel in this protein, and as they do so they spin part of it that looks like a tiny molecular water wheel, and as this happens ADP and a phosphorus ion are combined to form a molecule of ATP. Ten protons pushed into the intermembrane space by the electron transfer chain produce 2.5 molecules of ATP. The ATP is then used to fuel the creation of new proteins and to create gradients of other compounds, including sodium and potassium ions, across membranes.

Once again, the structure of a protein, this time ATP synthase, is critical for life. The spinning water wheel that is at the heart of the protein, is a structure central to making ATP, the molecule that powers life. Life is about building stable molecular structures that facilitate chemical reactions, including metabolism, that enable DNA replication. The DNA being replicated is itself structured in a way such that it contains instructions on how to assemble the molecular structures central for life.

Put another way, you are running a metabolism that is powering protein production, with the ultimate goal of making chromosomes to pass on to the next generation. Life took about 4 billion years to come up with you and me, and over those 4 billion years the most complicated life forms have progressively got ever more complex. Molecular structures have become more diverse and complicated, some cells have also become increasingly more structured, and cells have evolved to work together to produce structures such as bones, kidneys, tree trunks and fungal bodies. Life builds structures from proteins to brains that are sufficiently resilient that they can last a lifetime.

More complex structures required larger and more complex genomes. The self-assembly manuals acquired new genes, new switches, and the bodies that carried the genes got new cell types. I find it remarkable that evolution has self-assembled a self-assembly manual to produce intelligent organisms that have the wherewithal to understand not only what self-assembly manuals are, but also how they might have come about. This knowledge, which is uncontentious in the scientific community, has been arrived at via the application of the scientific method. When I briefly summarize the state of scientific knowledge, as I do throughout this book, I am not giving credit to the enormous effort that has been expended by scientists to gain insight. Thousands of scientists have taken observations, posed hypotheses, designed experiments to test them, analysed data and rerun experiments conducted by other scientists to make sure their findings can be replicated. The business of conducting science is time-consuming, and the generation of new knowledge is often stressful and challenging. Scientists are as human as everyone else, and the scientific endeavour is as frustrating as any other aspect of life. Key to conducting science is communicating about the work you have done, what you have found out, and how it might advance understanding. Before I turn to how life got started and began its journey to complex beings like you and me, I will briefly explain how scientists communicate their work, and how scientific consensus emerges. This will help explain why the astonishing complexity I describe above is accepted scientific knowledge.

Part of being a scientist is being able to effectively communicate. As a practising scientist I need to convince my peers that the novel research I do is robust. Scientists apply the scientific method to test a hypothesis, and then they need to tell other scientists about it. They do this through scientific papers that are written in a very formulaic manner. If you ever suffer from insomnia, go to Google Scholar, type in a scientific term or two and read the papers that are returned. You will come across dry prose such as:

We derive an equation to exactly decompose change in the mean value of a phenotypic trait into contributions from fluctuations in the demographic structure and age-specific viability selection, fertility selection, phenotypic plasticity, and differences between offspring and parental trait values. We treat fitness as a sum of its components rather than as a scalar and explicitly consider age structure by focusing on short time steps, which are appropriate for describing phenotypic change in species with overlapping generations.

This is from one of my more accessible papers. No wonder many non-scientists are put off by science.

When a piece of research is completed, it is written up and submitted to a scientific journal, where it is assessed for publication. An editor at the journal will initially read the submitted manuscript, and if he or she feels it might be well aligned with the journal's objectives, it will be sent for in-depth peer review. Between one and ten experts will then review the paper, sending their critiques back to the editor. The role of the reviewer is to comment on how novel the work is, report any flaws in the collection of data, experimental design, statistical analysis or modelling, flag any relevant literature that has not been cited and highlight any parts of the writing that are unclear. Their ultimate aim is to help improve the science, although some reviewers occasionally go further, choosing to humiliate authors in an attempt to destroy their confidence and possibly their careers. Mercifully, most reviewers are much more professional than this, providing helpful comments, although in my fifteen years of editing two of the leading journals in my field I have read a dozen or so staggeringly aggressive and unnecessarily hostile reviews. I would always rescind such reviews and strike the offending academic from the database of prospective reviewers. As reviewing is usually provided as an unpaid service, this is not a particularly harsh punishment, so I always hoped the anger the reviewer would doubtless feel when the paper was published would act as some form of reprimand for their deeply unprofessional behaviour.

Once a paper is published, other scientists will attempt to replicate results independently, particularly if the results are surprising or challenge accepted wisdom. If results can be repeatedly replicated, then the weight of evidence in support of the findings increases, eventually becoming accepted by the scientific community. Darwin's theory of evolution was contentious when published, despite Darwin's diligence in compiling large amounts of supporting evidence. In the 165 years since *On the Origin of Species* was published, biologists have worked out how to measure evolution, have developed models to predict it, and have studied it in laboratory and field experiments. They have collated vast quantities of evidence in support of evolution by natural selection. It is as real as gravity and electromagnetism. Scientists have repeatedly challenged the theory and, like Einstein's theory of general relativity, it has successfully withstood these challenges. Science progresses knowledge because the scientific method is concerned with challenge and replication, and together these make powerful bedfellows. Incorrect hypotheses and ideas are weeded out, with correct ones being supported with ever-increasing amounts of evidence until, like gravity and evolution, they become fact.

Scientific writing is dry, exact and formulaic, with some reviewers acting as the fun police, stripping papers of anything

that could draw a smile. I have even had the acknowledge-
ments section of one manuscript edited to remove thanks to
colleagues for uninformed debate. I was trying to be funny,
but the thanks were heartfelt, as the hours spent discussing the
work with colleagues with expertise in other areas helped me
identify which parts of the problem were hard to grasp by the
non-specialist. Nonetheless, the occasional author sneaks wit
past humourless reviewers and editors. One of my favourites
is a 1974 publication in the *Journal of Applied Behavior Analysis*.
Dennis Upper's publication 'The Unsuccessful Self-Treatment
of a Case of "Writer's Block"' is nothing more than a blank
page. A footnote by reviewer A in part states:

> I have studied this manuscript very carefully with lemon
> juice and X-rays and have not detected a single flaw in either
> design or writing style. I suggest it be published without
> revision. Clearly it is the most concise manuscript I have
> ever seen – yet it contains sufficient detail to allow other
> investigators to replicate Dr Upper's failure.

Papers have become the currency of science, with the
mantra 'publish or perish' being used increasingly over the
last two decades. The large numbers of applications for some
jobs – sometimes many hundreds for a single position – mean
busy members of appointment panels produce lists of competitive applicants by looking at publication records rather
than by reading the papers, and in response to this early-career
researchers seeking a faculty position in a university rush to
publish. Over the years, many of my graduate students have
asked me how many papers they need to publish to get an academic position. Coupled with this, in the UK, the government
determines funding to departments and universities by an

assessment exercise based partly on publication outputs. With the public rightly demanding that their taxes be appropriately spent, governments wanted a way to justify how higher education funding was spent. Papers provide a simple metric. The problem with the papers-as-currency landscape is that good science often takes time, and this means that many papers published today say little. In addition, many of the top journals in the field, where scientists like to publish, favour papers reporting exciting new results rather than replicating existing ones. Scientific culture, at least in my field in the UK, means that the scientific method is not being used as effectively as it should be. Funding needs to be available for studies to attempt to replicate published results, and the best journals should publish results from such studies. The focus on papers is sensible, but it does mean that books count for little in the government assessment of university performance or in the scientific job market, and writing books has become a lost art in many university science departments.

I am extraordinarily lucky to have a permanent job at a university that is less obsessed with publication metrics than many. I am appraised by my line manager once every five years, and I am expected to try to produce a few really good publications over that time rather than large numbers of papers that have limited impact. I am also encouraged to participate in promoting science to the public. This enlightened attitude provided me with the opportunity to write this book, and, although my life's aim since my brush with malaria had been to understand why I existed, I only decided to try to write what I had learned and concluded in a popular book when I moved to Oxford in 2013. I knew I needed a year of largely uninterrupted time to break the back of my book-writing project, and I planned for 2020 to be that year, as by that time Sonya and I would be due a sabbatical. The pandemic intervened, so we had to delay our sabbatical by a year, but in September 2021 Sonya and I headed to the University of Queensland in Brisbane and I began to write.

I did everything rather back to front. I didn't know what the fate of the book might be when I started writing it, and my initial aim was to write it for my children. However, when I showed draft chapters to my youngest two, Luke and Georgia, they told me it wasn't for them, and they wouldn't read it. I admit I was a bit dispirited. Nonetheless, colleagues and friends in Brisbane were more positive, saying they liked the idea of the book and they would love to read it, so I persevered, and as I neared the end of the first draft I started to think I might be able to publish it. A bit of investigation revealed I had three options. I could self-publish, and if I did this, I would be responsible for marketing the book and would likely only sell a few tens or hundreds of copies. Alternatively, I could work through one of the academic publishing houses and would be writing for scientists, and this became my fallback option. I'd need to strip all personal anecdote, humour and any hint of elegant prose from the text, but one of the university presses would deal with the printing and market it to an academic audience. Alternatively, I could work with a literary agent who would try to find me a publisher of popular science books and it might be more broadly read. Those in the know told me this was hard, and that many academics failed to be recruited by an agent, but given I'd written the text with my well-educated but non-academic grown-up kids as the intended audience, I decided to give it a go.

The first two agents I wrote to never replied, but the third did. I met up with her, she signed me up as an author to her literary agency, and I was over the moon. She explained that it was unusual to have a full draft of a book, and that I would need to write a proposal to pitch the book to a publisher. I did just that, and an editor who worked for Penguin Michael Joseph signed me up.

I knew what would happen next, for it had happened at every stage of my career. That night was sleepless, I felt that I was a fraud and that this shouldn't be happening to me. My agent and my publisher had both made a dreadful call, and I was way out of my depth. I had had the same feeling when I won a scholarship to conduct graduate studies, when I was offered my first faculty position, when I was awarded my first research grant, when I was promoted to Professor, when I was appointed to Oxford and when the faculty of my department voted for me to be their head. But I knew what I had to do. I took Woofler for a long walk, and logically thought it through. My agent and my editor knew what they were doing. Both had worked with some great authors, and they were at the top of their profession. They had faith in me, so, although I still had my doubts, I owed it to them to produce the best book I could, as they were working hard to achieve the same outcome as I was – an informative and enjoyable book. I would be doing them a huge disservice if I were to continue to consider myself as an imposter. I was also determined to enjoy the experience, as I also believe I will only live once, so I should make the whole experience fun. I now get a kick from dropping into conversation that I have a literary agent and an editor and seeing the looks of disbelief on the faces of my colleagues, and after my editor told me my dog was soon to be famous, I ascribed a metric of success as the number of pets called Woofler I meet in the future. I assume those pets will be dogs, for what fool would call their cat Woofler? For that matter, Woofler is a pretty daft name for someone to give to a dog.

Although you may now consider me a bit of a simpleton for naming a dog Woofler, I am, like you, staggeringly complex. Or at least the chemistry that keeps me alive is. Prior to the brief scientific publishing detour, you will have come to appreciate quite how complicated we all are, and you will likely be wondering how life self-assembled the first self-assembly manual. We do not know for certain, but we have some ideas and mounting evidence, and I suspect it won't be too long before scientists produce simple life in the laboratory from a cocktail of simpler molecules. Part of the trick is working out what those simpler molecules would have been on our young planet, along with the conditions required for them to assemble into the first living organism.

I'm going to assume that life began on Earth rather than in space. Even if life didn't start on Earth, it is usually assumed to have begun on a planet or meteoroid rather than in space itself. A few researchers have championed a hypothesis called panspermia whereby life started on another planet before colonizing Earth. Panspermia does not solve the challenge of how life got started, but it does provide for a wider range of environments beyond those found on early Earth for where it could have begun. A downside is that the theory requires not only an understanding of how life began elsewhere but also an explanation for how it travelled between planets. This additional challenge, along with the sparsity of knowledge about the chemistry of other planets, means few scientists actively study panspermia, and I will discuss it no further, thus assuming that life began on Earth. Despite this assumption, many organic compounds do form elsewhere in the galaxy, including in nebulae. They have been detected on meteoroids, found on meteorites, and their signatures observed on other planets and in other star systems through the deployment of high-tech telescopes, but these finds are of organic molecules rather than of replicating life. Life probably began on Earth but was given a helping hand with much of the raw material that the first cells used coming to our planet from space.

The formation of the Earth, and the subsequent settling of its orbit into the habitable zone 150 million kilometres from the sun, provided necessary conditions for the emergence of life. When you think of what you need to live, you will probably think of oxygen, because without it you would die. Early life did not rely on oxygen, and it would probably have killed it. The atmosphere of early Earth during the period when the first life arose was very different from the air we breathe today. The first thick atmosphere was created by the collision of Theia with Earth, and scientists think it consisted primarily of nitrogen, carbon dioxide, water vapour and sulphur dioxide, with oxygen being nearly entirely absent. The surface of Earth in its youth was hot, and the first atmosphere had a high concentration of greenhouses gases, and this meant it took time for the Earth's surface to cool to a temperature where water could occur as a liquid – the first key requirement of life. By 3.9 billion years ago, the Earth had cooled sufficiently for this to happen. The first life is thought to have evolved quite quickly once liquid water was available, which means that the climate of the early Earth when life formed was much hotter than it is now. Our planet's average temperature would have been close to the boiling point of water, compared to the chilly 13.9 degrees Celsius of today.

A second key requirement for the emergence of life was the availability of carbon-based organic molecules. During the formation of the Earth from stellar dust particles, some organic compounds used by life would have been present. However, these would have been destroyed by the immense energy created by the planetary collision that led to the formation of the moon, as the collision turned the surface of our young planet into a sea of molten rock that would have ripped apart even the hardiest organic compounds. Molecules that were necessary for the first life were consequently either synthesized on Earth after the collision that formed the moon, or were formed in space before arriving on meteoroids, asteroids and dust particles entering our planet's early atmosphere.

Evidence of complex organic compounds being generated in space comes from the examination of meteorites, with the most studied of these having crashed into Earth in late September 1969. The meteorite's impact was observed close to Murchison, Victoria, Australia, and the meteorite was quickly collected before it could become contaminated by molecules from Earth. The Murchison meteorite contained seventy different amino acids (life uses twenty) among the 14,000 molecular compounds so far identified. The Murray and Tagish Lake meteorites have also been studied, but not in quite as much detail, and nucleobases have been discovered on all three rocks. In 2019, scientists even discovered sugars, including ribose, on meteorites. Space rocks can be rich in organic compounds, and some scientists have predicted that the Murchison meteorite contains millions of distinct organic compounds. The Murchison meteorite is also old, and much older than the Earth. Dating of silicon carbide particles found within it revealed it could have formed about 7 billion years ago. What this means is that complex organic compounds, including those life uses to build DNA, RNA and proteins, appear to be easily formed in space and can fall to Earth on particles of space dust or meteorites.

Phosphorus is another element that is key to life, being an ingredient of both DNA and RNA. Although phosphorus is a common element, much of it would have been rather inaccessible to early life, being locked up in rocks and minerals. Life would have needed more easily accessed phosphorus-containing molecules. Some may have arrived on the meteorites that brought water and organic compounds to Earth, but another possibility is that lightning strikes may have also played a role. The atmosphere on Earth around the time life emerged would have featured frequent lightning strikes, and laboratory experiments reveal these could have helped release inaccessible phosphorus, making it available for life in the form of molecules called phosphides, phosphites and hydrophosphites. I suspect that both lightning and meteorites played a role in providing accessible phosphorus for the first organisms, but this key molecule would have been present on the early Earth in a form that life could use.

Your cells have high concentrations of complex organic molecules that contain elements like phosphorus, carbon, oxygen and nitrogen. The chemistry of organic molecules that happens within a living modern-day organism such as you, me, an earthworm, or even a bacterium, is very different from the chemistry that happens in the external environment of today's Earth. Within each of your cells you are continuously making large, complicated molecules such as proteins. Outside of your body, on the floor, or the table, the chemistry that is occurring is different. The chemical activity in the external environment is much, much simpler than that which is happening in your cells, and it does not involve large organic molecules to anywhere near the same extent.

Scientists describe the external inorganic environment as being at an equilibrium, with your cells being in a state that remains quite stable while you are alive, but which is very different from the external equilibrium of the environment outside of you. The difference between the steady state of cells in living organisms and the equilibrium state of the external environment can be thought of as a distance. The greater the difference between the chemistry in your cells and the chemistry in the external environment, the greater the distance. The distances observed between modern-day cells and the external environment in which they find themselves are much larger than the distances would have been when the first wisps of life emerged. It is physically and chemically impossible to instantaneously create a cell that is as different from the external environment as the cells in you and me are from the outside, non-living bits of our planet. Given this, early life must have evolved within a non-living environment that was rich in the compounds it needed, and as life became more efficient over time, it became better at commandeering these compounds, leading to a greater difference between the chemistry going on in the inside and on the outside of life's early cells. As life got older, the distance between the chemistry going on within cells and within the external environment would have increased. The next challenge in understanding how life emerged is to understand how the organic molecules, water and phosphorus molecules from which the first life emerged become sufficiently concentrated that a simple cell might arise.

There are several competing hypotheses as to where

organic molecules could be sufficiently concentrated for early life to emerge, with volcanoes playing a role in the two that are most strongly favoured by scientists. Life is thought to have evolved in either volcanic ocean vents or in freshwater hydrothermal fields like those found in Yellowstone, Iceland and Japan today. In volcanic oceanic vents, organic compounds are hypothesized to have become concentrated in tiny pores in volcanic rocks. In contrast, in hydrothermal fields on land, repeated wetting and drying events caused by rain rich in organic compounds followed by evaporation or large tides caused by the moon are argued to have created small pools suitable for life to emerge. Currently it appears that life could potentially have evolved in either freshwater or saltwater volcanic environments, with chemists yet to successfully rule either out. As we learn more about complex organic chemistry, emergence-of-life researchers may eventually conclude that life could only arise in one of these environments. Regardless of which environment it arose in, the next step is the start of a complex chemical reaction that is called autocatalytic.

An autocatalytic reaction involves a form of replication. A very simple example of an autocatalytic reaction is adding two compounds – for simplicity, let's call them A and B – into a beaker in equal amounts, letting the reaction happen and observing only B at the end of it. B is called an autocatalyst, and in this case one molecule of B can convert one molecule of A into another molecule of B. A catalyst is a compound that enables a reaction, and adding 'auto' in front of the word means B makes copies of itself if it has the right material to do so – in this case molecules of compound A. We start with one A and one B molecule and end up with two Bs. Perhaps there is some waste product too. The reaction may occur giving out heat, or it may require energy to proceed. Which will depend on the chemical properties of A and B.

Life is a very complicated form of an autocatalytic reaction. For example, we could replace molecule A with prey and molecule B with a predator. A lion eats zebras and uses the energy from the zebra's flesh and bones to produce lion cubs. Lions will continue to do this until zebras (or other prey) become scarce. I am not the first person to notice this: the equations that chemists use to consider one type of autocatalytic reaction are identical to the equations that ecologists use to describe how predators and prey interact.

We don't know what the autocatalytic reaction was that was the key to life getting started. Perhaps it involved a number of chemical steps which resulted in molecules of nucleobases and amino acids making copies of themselves, a bit like DNA and RNA do today, but in a simpler way. Some scientists have argued that RNA evolved before DNA, and that RNA was involved in an early form of autocatalytic reaction. The RNA-first hypothesis, as it is called, is attractive in that RNA encodes information much as DNA does, but it is more reactive and can act as a catalyst in a way that DNA cannot. DNA is stable and non-reactive. Arguments against the RNA-first world are that each RNA molecule is not particularly stable, and replication errors may have been too high for autocatalytic reactions involving it to persist for long. Regardless of the details of the first autocatalytic reaction, it would have had three properties. First, some of the chemicals involved would have become locally abundant. Second, at least one of the chemicals involved could have taken a variety of forms, mutating from one form to another – replication would not have always been perfect. Third, the various forms would have competed against one another, with the superior form winning.

Any compound that can make copies of itself can spread fast. Autocatalytic reactions repeatedly double the number of self-replicating molecules, as long as conditions allow. One molecule makes two, two molecules make four, and so on. The astonishing power of such exponential growth is hard to comprehend, but the numerical consequences of continuing to double numbers is nicely illustrated by an old Indian proverb.

A rich king of old was a keen and competent chess player who liked to play strangers he encountered during his travels. One day he encountered a sage who had played chess all his life. The king challenged the sage to a game, and to motivate him he offered him a prize. The sage said that if he won, he would like the king to put one grain of rice in one of the corner squares on the chessboard, two grains on the next square, four grains on the next, and on each new square to double the number of grains of rice that were on the previous one until all sixty-four squares were full. The king agreed, and subsequently lost the game.

When trying to pay the sage his due, the king quickly realized his folly. By square 21, the king had already handed the sage over a million grains of rice. By square 64, the king would have needed over 200 billion tons of rice, covering all of India in a 1-metre-thick ricey blanket. Quite why the sage wanted so much rice is not explained, but the proverb shows how powerful a force exponential growth can be. In just ten cycles of an autocatalytic reaction, one molecule can produce 512 copies of itself, with over a million copies produced by cycle 20. Such staggering growth cannot continue indefinitely. It ceases when any required resource becomes unavailable. In an autocatalytic chemical reaction, a key chemical will become used up.

The first autocatalytic chemical reaction on Earth that was the precursor to life was not itself alive. It had to become more complex, and involve membranes and other compounds before it could be counted as living. In order to take these steps, imperfect replication was required. The reaction did not always result in identical copies of key molecules being made all the time. On occasions the new molecules being formed would be imperfect copies of the parent molecule. Some of these new versions would have been better autocatalysts than their predecessors, being more effective at commandeering the chemicals required to run the reaction, being able to use a new chemical or energy source, or being more energy-efficient. Each new autocatalyst that was better than its less effective ancestor would spread faster, outcompeting it and eventually driving it extinct. The generation of new variants through imperfect replication, followed by competition for resources, is the basis of evolution. Evolution works for autocatalytic chemicals as well as living organisms. It also works for viruses. Most scientists do not consider viruses as being alive, but they are able to invade living cells to make copies of themselves.

We all became familiar with imperfect replication, exponential growth and the power of competition during the COVID pandemic. Despite politically motivated statements by parts of the US government and conspiracy theorists, Sars-Cov-2, the virus that causes COVID, originated in wild animals, and most likely in bats. The first variant that emerged in China was called the L-strain, and this mutated to form the S, V and G strains. A mutation occurs when the genetic code is not properly replicated such that the genome of the initial virus particle, or virion, differs from that of the copy. Most of us paid little attention to these mutations, but in September 2020 we all listened when scientists announced the alpha variant that arose in the UK and quickly started spreading around the world. The Beta and Gamma variants subsequently emerged in South Africa and Brazil, but they made few ripples. In 2021 the Delta variant arose in India and spread across the globe, replacing other variants. In turn Delta was replaced by Omicron variants.

Viruses consist of genetic code wrapped up in a protein coat. Sars-Cov-2's genetic code is written in RNA, and the protein coat includes a spike that is key to the virus infecting our cells. When a virion breaks through a host's cell membranes, the RNA uses the cell's machinery to make new copies of the virus. As viruses make copies of themselves, they often make the host ill and sometimes even kill it. Humans became a host – a resource – for Sars-Cov-2, and each new variant arose from an existing virus through imperfect replication. Mutations to the virus's genetic code occasionally occurred when the host cell failed to make a perfect copy. Some of these mutations resulted in new spike proteins, the part of the virus that helped it enter host cells, and some of these new spike proteins made the virus better at infecting new host cells. Other mutations would have made the virus useless and unable to infect new host cells, but these strains were destined to become rapidly extinct and we knew nothing of them.

Each successful strain was replaced by a new one that could spread around the globe because it was better at infecting people than the last strain. Humans were the resource these variants were fighting over, and the Omicron variants were best at being transmitted between people and at making copies of themselves. Omicron, Delta, Beta and Alpha all grew exponentially, but Omicron was able to do it fastest because it was better at infecting people. It commandeered its resource – us – at a faster rate than its less competitive ancestors. Early autocatalytic compounds on the early Earth would have done the same thing with the compounds they needed to make copies of themselves. The species of molecules that could most rapidly use available chemicals to make copies of themselves would displace the less competitive types. Competition coupled with accurate, but occasionally imperfect replication means the most efficient type wins out.

Most research on autocatalytic reactions focuses on relatively simple cases involving only a few chemicals involved in a cyclic chain. The reactions that began in the environment where life arose were probably more complicated than this. I imagine a soup of large numbers of different, complex organic molecules concentrated in either a marine or freshwater environment. In this soup, networks of reactions took place. More specifically, there would have been an autocatalytic reaction involving a handful of chemicals. We can think of these reactions as being analogous to the replicating germline of modern-day life. However, other chemical products produced by this core reaction could have facilitated the autocatalytic reaction, playing a role analogous to that of the disposable soma. If this speculation is correct, separation between the germline and the disposable soma may have been a feature of precursors of early life. However, for such a complex network of reactions built around an autocatalytic core to become life, it would need to be enclosed in a membrane. Where did membranes come from?

The membranes that life uses are made of a type of

molecule called a lipid, and lipids are made up, in part, of smaller components called fatty acids. You probably won't be surprised that fatty acids have been found on meteorites, including the one that hit the ground near Murchison. Like all other key components of life, these essential molecules are not restricted to Earth, forming elsewhere in the solar system too.

The fatty acid molecules are long chains built from carbon, hydrogen and oxygen atoms. In cell membranes, two fatty acids are joined with a molecule of a compound called glycerol and are then attached to a phosphate molecule. Phospholipids, as the resulting molecules are called, look a bit like a tadpole with two tails. The head is the phosphate molecule, and the tails are the fatty acids. When placed in water, these phospholipids spontaneously form two joined layers, a so-called lipid bilayer, with the heads on the outsides of each layer, and the fatty acid tails pointing inwards towards one another. These bilayers form because phospholipids will not dissolve in water. Instead, different parts of the molecule have different electric charges, which means they interact with water molecules and with one another to form a bilayer. The phosphorus ions in the head of the molecules are attracted to the negative-charged parts of the water molecules. In contrast, the fatty acid tails are hydrophobic – meaning they are repelled by water. These properties, stemming from the electromagnetic force, mean that the molecules rapidly form bilayers in water as their heads and tails are respectively attracted to and repulsed by the water molecules. It is easy to form phospholipids. When placed in water, these lipids spontaneously form membranes, and it is these lipid bilayers that are the basis of membranes that surround cells.

Once a lipid bilayer forms it acts as a barrier to ions,

proteins and other molecules. Such molecules cannot cross the bilayer as it acts like a fence around a field that keeps sheep trapped inside and wildlife out. One challenge early life would have faced was how to move molecules across the bilayer, and many researchers argue for leaky membranes. Modern life solves it with proteins embedded in the membranes that surround cells, as in the mitochondria example described earlier. These embedded proteins act a bit like gates in my field-of-sheep analogy.

Exactly how autocatalytic reactions became embedded within the first cell, or how the first cells divided, is unclear. Nonetheless, the necessary molecules for life were abundant on the early Earth, and once autocatalytic reactions began they would have spread. Membranes form easily, so the key component of a replicating cell would have been available for chemistry to work its magic. But there is still one thing we have not touched on. Where did the energy to run these reactions come from, and how did life use it? This is where volcanic activity comes to the fore, potentially driving the metabolism that early life would have used.

Life today is powered either by energy from light, or by energy from highly reactive molecules such as hydrogen sulfide or, in the case of animals, by energy gained from consuming other forms of life. The chemical reactions involved in metabolism differ depending upon the source of energy, but the result is always the same: the creation of molecules of ATP. ATP is life's fuel. It is reasonably stable but can be coerced to release energy easily enough, and it does not blow up. It is a much better chemical to power the germ cells and the disposable soma than azidoazide azide.

Life needs energy to run, but in doing so also gives off energy. That is why you are warm. The travails of staying alive require you to burn energy. Light is extremely abundant on Earth, and anywhere close to stars, and the cyanobacteria that formed the first stromatolites ran their metabolism using light. However, scientists think that the first living organisms used chemosynthesis to power their replication. They used reactive chemicals, and a great source of these would have been volcanic vents in the ocean or on land. Volcanoes produce large amounts of compounds that readily share, donate or steal electrons from other compounds. In places like Yellowstone, bacteria living in sulphur-rich hot pools run their metabolism from hydrogen, hydrogen sulfide and carbon dioxide, and that is a good candidate suite of chemicals to power the first cells.

The use of reactive chemicals by organisms to power metabolism is known as chemosynthesis. Modern-day bacteria use a range of chemicals, but each produces a sugar from a carbon-containing molecule such as carbon dioxide or methane gas. Some chemosynthetic bacteria use hydrogen to power the production of sugar, while others use ammonia or hydrogen sulfide. Organisms that require oxygen to live use a different type of metabolism called the Krebs cycle, or citric acid cycle, which you might remember from school. Although involving more reactions than most forms of chemosynthesis to create sugars, key components of the Krebs cycle naturally emerge in the right mix of chemicals. Like the formation of cell membranes, various forms of metabolism naturally occur under the right conditions.

In the next chapter, we will meet an ancient primitive organism called LUCA from which everything on Earth today is descended. Although long extinct, scientists have had a stab at reconstructing its genetic code, and this reconstruction points to genes involved in the chemosynthesis of hydrogen sulfide or similarly reactive chemicals. I don't know whether the first cell used ATP, but LUCA may well have done, suggesting that the universal fuel of life today was adopted early. Scientists have yet to find ATP on meteorites, but they have found the separate adenine and triphosphate building blocks. Like other key molecules of life, its building blocks can be made elsewhere in the solar system before being delivered to Earth. Emergence-of-life chemists have also made compelling arguments that proton pumps have been central to life from its earliest days, so ATP may have been used surprisingly early.

The details of exactly how and where life began are unknown, and there is still a lot to discover, but scientists have made much progress. We know why life emerged – like all chemical reactions, it happened because it was energetically the easiest option. Given the right mix of chemicals and energy inputs, life is the favoured outcome. We also have some idea what had to happen for life to start. Scientists know that autocatalytic reactions fuelled by a metabolism run on chemicals like hydrogen sulfide became enclosed in a membrane, and that this happened in a solution rich in organic compounds both seeded from space and produced on Earth. As early life continued to develop, driven by errors in replication, the chemistry inside the membrane became more complicated than on its outside. We also know that early life spread because the exponential growth of a molecule is a powerful force. Competition led to increases in complexity until the first protocell existed, but we do not know the details of each step. In time, DNA and the genetic code life uses today became established, and, from then on, life did not look back. There is a huge divide between the simple chemistry we conduct in labs and chemistry of even the simplest life, but we have a roadmap of what had to happen, and we know that the key molecules would have been available on the early Earth. What we don't know is how it all happened, and what the right mix of chemicals is, how much variation there might be in that mix, or how long it might take. For this reason, the emergence of life is still shrouded in mystery, but the shroud is steadily being lifted, and in the next decade or two I suspect life's emergence will become increasingly better understood, and may even be replicated in the lab.

My conclusion for how life emerged after reading extensively on the topic is that pools rich in amino acids, phosphates, lipids, nucleobases and other key building blocks of life formed in freshwater environments in hydrothermal fields, although I cannot rule out life emerging in the oceans. Hot water temperatures, coupled with a steady influx of chemicals from volcanic activity, resulted in stable, complex, organic compounds such as DNA, RNA, phospholipids, proteins and ATP forming. These large and complex molecules formed because they were energetically favoured, being the stable state of the system. The structure of some of these molecules facilitated metabolism and replication. Life assembled itself, with its autocatalytic abilities and complex structure emerging from a rich soup of complicated organic molecules that were themselves formed from a concentrated pool of simpler chemical building blocks either forged on Earth or delivered to our planet on meteorites. The problem in testing this hypothesis is we don't know enough about the environment in which life emerged.

If scientists were to simply take random mixes of chemicals and subject them to the environment thought to exist on the early Earth it may take millions of years before life arises in a test tube. Although life emerged quite quickly on the early Earth, it is only quick on the scale of the 4.5 billion years of Earth's history. It may have taken millions, or even hundreds of millions, of years to appear. Fortunately, chemists can do much more than randomly mix chemicals together. Through the study of meteoroids and meteorites they are developing an understanding of which building blocks were likely available to early life. Computer simulations of the behaviour of these compounds built from our understanding of the workings of electromagnetic force are playing an increasingly important role in emergence-of-life research, and it strikes me that identifying how life emerged is a problem that may be tractable for the latest artificial intelligence algorithms. AI has already helped gain insight into how chains of amino acids fold into proteins, a problem that humans had long found intractable. AI can be very effectively applied to help us understand complicated chemistry. The irony of using non-living artificial intelligence to help create the living is not lost on me. It would grab headlines around the world.

Once evolution had moulded the DNA and the genetic code that life uses today, things really took off, and life proliferated. Life has clung to Earth for nearly 4 billion years and has coped with everything the universe has thrown at it, from solar flares to meteor impacts to frigid temperatures. It has not just coped, it has thrived, and now it occupies nearly every crook and crevice of our planet, from the floors of the deepest oceans to the skies above the tallest mountains. The next part of our history focuses on how life spread and developed from its simple origins, and how it overcame some very significant challenges.

Life Conquers All

In the previous chapter I described how and why competition between autocatalytic molecules would have resulted in the evolution of replicating systems of chemicals in the environments where the first life formed. I doubt the description upset many readers. In contrast, when evolution is discussed with respect to more complex forms of life, and particularly humans, some people challenge its primacy in determining our existence. Yet the evolutionary processes that produced the first living organism are the same as those that produced humans nearly 4 billion years later. Evolution is as real as gravity, electromagnetism and the strong and weak nuclear forces, and without it we could not exist. Biologists understand evolution to about the same extent that physicists understand the strong and weak nuclear forces and chemists understand electromagnetism. They know how and why evolution creates new species. Evolutionary biologists like me have a clear definition of evolution and we know how to measure it. We understand how genetic differences arise and how natural selection operates on this variation to generate new forms of life. We understand how changing environments result in evolution, and in some cases biologists can predict how natural selection will change animals and plants. If you accept science as a way of finding out how and why the universe works as it does, you should accept that evolution is as real as the four fundamental physical forces discussed earlier.

There are two reasons why evolution can be a challenging topic. First, some people deny evolution because they find it hard to accept that humans evolved from simpler life forms such as single-celled bacteria and, more latterly, chimpanzees. Yet the comparisons of genomes between species as diverse as bacteria that cause disease, bananas, sea squirts, worms, fruit flies, dogs, gorillas and us provide overwhelming proof that all forms of life on Earth are related to one another. To generate this astonishing biological diversity, evolution needed billions of years, and during that time it invented death and sex, tiny structures in cells called mitochondria and chloroplasts, thousands of new types of molecules including forms of carbohydrate, protein and lipid, numerous cell types, and multicellular organisms such as the silver birch tree, Woofler and you. This chapter is about what had to happen for animal, plant and fungi species to become complicated. There are thought to be about 8.7 million species of plants and animals alive on Earth today, along with an unknown number of fungi species, and hundreds of millions of species of single-celled microbes. Evolution has invented a vast number of ways for organisms to make a living. We will encounter some of these as we think about what happened for the first mammals to evolve.

The second reason that evolution can be challenging is it is a quite complicated concept. Some readers will find it harder to understand than the workings of the four fundamental forces. Evolution is difficult because it involves several things happening at once. On their own, each of these things is easy to grasp, but when combined things can become complicated. I shall try to briefly summarize.

All organisms need resources such as food, water and shelter to live, and sexually reproducing species such as humans need to find a mate to reproduce. However, there are many things in the environment such as predators, or disease, or large numbers of competitors, that can make finding resources, including mates, difficult. Individuals that are good at finding, acquiring and using resources while avoiding the threats that the environment imposes upon them tend to have particular attributes such as the ability to deter things from eating them or the ability to mount an immune response to an infection. These attributes are determined, at least in part, by an individual's genes – its DNA code. Those individuals that succeed in acquiring resources and avoiding threats tend to produce the most offspring, and this means their genes are represented in more individuals in the offspring generation compared to individuals who did not have these particular attributes. These genes are likely to produce the desirable attributes in offspring that made their parents successful. Evolution proceeds as genes that determine attributes that enable individuals to acquire resources while avoiding threats increase in frequency within a population. Over short periods of a few generations, evolution happens quite slowly and not a huge amount of change is observed. Over many hundreds of thousands or even millions of generations, populations can significantly genetically diverge, and new species can evolve. Evolution is consequently a process driven by which individuals are best at surviving and reproducing, but evolutionary change over a few generations is usually quantified by looking at how populations change over time or by comparing genetic differences between separate populations of the same species. On longer timescales, evolution is quantified by looking at genetic differences between individuals of different species. And there are a lot of species to compare.

When my children were young and I asked them to name their favourite species they would say lion, dolphin or dog. I failed to persuade them of the merits of species like *Candidatus Prometheoarchaeum syntrophicum* strain MK-D1, a species of single-celled organism found in the depths of the ocean. Each new MK-D1 cell starts life as a blob and develops by growing tentacle-like projections. These thin structures search out bacteria that acquire their energy from hydrogen molecules, and having found them, MK-D1 then gets the energy to power its metabolism from their waste products. MK-D1 lives off the poo of species of bacteria that do not use oxygen, a gas we often think of as essential for life.

Strain MK-D1 is a form of life you might be unfamiliar with. Biologists divide all life on Earth into three domains called the Bacteria, the Archaea and the Eukarya. Bacteria and archaea are always single-celled organisms, and these cells are relatively simple. In contrast, eukaryotes have more complicated cells, subdivided by internal membranes. Some species, like yeast, are single-celled organisms, but eukaryotes can also be multicellular species such as seaweed, fungi, trees and animals. Each of these three domains of life can be subdivided again and again, into hierarchical levels of organization called kingdoms, phyla, classes, orders, families, genera and species, until each species is eventually classified. I remember this with the pneumonic 'Do Keep Pigs Clean Or Farm Gets Smelly' with the D in Do representing domain. Our full classification from the species up is *Homo sapiens*, in the genus *Homo*, which sits in the Hominidae family, which in turn is classified as being in the Primate order, in the Mammalia class, in the Chordata phylum, in the Animalia kingdom and, finally, in the Eukarya domain. In contrast, the microbe species *Escherichia coli* that can sometimes make you sick is classified as: genus *Escherichia*, family Enterobacteriaceae, order Enterobacterales, class Gammaproteobacteria, phylum Pseudomonadota, and domain bacteria. I'll spare you the full classification of strain MK-D1, simply stating that it is a member of the Archaea domain, a group of single-celled organisms that biologists once thought were bacteria but have more recently realized are not, and are instead more closely related to animals, plants and fungi.

Species in the same genus are closely related to one another, typically having shared a common ancestor within the last few million years. Humans and chimpanzees are not in the same genus but they really should be, and our common ancestor lived between about 6 and 10 million years ago. As we move up through the classification system to family, order, class, phyla, kingdom and domain the species we encounter become progressively less closely related to us. MKD1 and humans are in different domains, and we shared a common ancestor over 3 billion years ago.

The reason I like strain MK-D1 is not only because it lives an unusual life but because it does so very slowly. It takes almost a month for a single cell to grow and divide into two cells. You might think that a month isn't very long, but in the world of single-celled species it is an eternity. *E. coli*, for example, divides every twenty minutes. In the time it takes for one MK-D1 cell to become two individuals, if *E. coli* had access to unlimited food, one cell could grow to produce a colony of two raised to the power of 2,190 bacteria, a number so astronomically large that my computer returns infinity when I ask it to calculate it. Another example of the power of exponential growth.

Strain MK-D1 is also much more biologically interesting

than a lion, dolphin or dog. It has genes that, until its discovery, were thought to be found only in eukaryotes, leading scientists to hypothesize that MK-D1 can shed light on how the complicated cells that eukaryotes such as you and I are built from evolved from the much simpler cells of archaea. As an example, single-celled eukaryotes are able to do something called endocytosis, a feat beyond bacteria. Endocytosis is the process these species use to envelop small objects such as particles, viruses and sometimes even smaller cells in order to bring them across their cell membrane. It is a process that helps bring nutrients inside a cell for it to use. Cells do this by bending their membrane to form a bubble around the object to be brought into the cell, before pinching the bubble off to create a small membranous sphere in the cell's interior that contains the object. It is the cellular equivalent of you swallowing a morsel of food. MK-D1 produces a protein called actin that is necessary for endocytosis, and which is found in all eukaryotes but in no bacteria. Strain MK-D1 provides insight into the common ancestor of archaea and eukaryotes by providing clues into how life made a major jump in complexity, and biologists think it may also help us understand how endocytosis evolved. However, gaining this knowledge was difficult because it is very hard to keep MK-D1 cultures alive outside of its ocean habitat. It is only in the last few years that a group of biologists in Japan succeeded in keeping a population of MK-D1 alive in their laboratory. Lions, dogs and dolphins are not only much easier to keep in captivity but could never have existed if species similar to MK-D1 had not evolved genes that produce actin, a gene central to all complex life.

Strain MK-D1, *E. coli* and you all use the same genetic code. We all use twenty amino acids to build proteins, and these amino acids are coded by the same set of nucleobase triplets. Thymine–thymine–thymine codes for the amino acid phenylalanine not only in these three species, but in all species of life on Earth. The genetic code is universal to life on our planet. Although every species uses the same genetic code, different species have different genetic sequences, allowing them to make different proteins. By comparing how similar genetic sequences are between species, biologists can work out how closely related different species are. Because humans and chimpanzees shared a common ancestor only a few million years ago, the genetic sequences of the two species are much more similar than the genetic sequences of humans and MK-D1.

Despite the ubiquity of the genetic code, the very earliest life would not have used it, but it didn't take long for life to adopt it. Inferring exactly what happened and when 4 billion years ago is not easy, but the genetic code we use today was probably in operation only a few million, or perhaps tens of millions, of years after life appeared on Earth. Although we do not know exactly how long ago the genetic code evolved, we do have a name for the organism that first used it: LUCA, or the Last Universal Common Ancestor, a term that scientists have used since the 1990s. You, me, the trees outside, the birds that sit in them, along with the bacteria in their guts, the invertebrates that parasitize them and the microscopic fungi that can cause them disease, are all descended from LUCA. There is a direct line of descent from LUCA to you, from LUCA to MK-D1, and from LUCA to *E. coli*. You might feel fortunate you have a direct line of descent from LUCA, but so too does every organism alive on Earth today, from each cockroach, to Woofler, to you. LUCA has very many and very diverse descendants. In this chapter, I explore how and why complex animals such as humans evolved. Why are we so different from dogs, *E. coli* and strain MK-D1? What had to happen in the 4 billion or so years since the genetic code evolved for our genetic self-assembly manuals to arise? Evolution is the process that has moulded life on Earth, and to understand it it is necessary to turn to the workings of DNA, phenotypic traits, and natural and sexual selection, terms I define in the coming pages.

The tree of life starts from LUCA and describes how new species evolved from existing ones. It is not complete, and there are many gaps, particularly the further back in time we go, because species alive in the distant past are extinct, and for many of them no fossils have been found from which they can be described. Nonetheless, it is clear from even the incomplete tree of life that different now-extinct species had very different fates. Some produced many descendant species that are still alive, while others left none. *Anchiornis huxleyi* was one of the first species of birdlike dinosaurs, evolving about 160 million years ago, and even though it is long extinct all of the ten thousand or more species of bird alive today are thought to be descended from it. *Tyrannosaurus Rex* is at the other extreme, leaving no descendant species, as its lineage went extinct following the meteor impact that brought the Cretaceous geological epoch to an end 66 million years ago.

We share nearly all of our line of descent from LUCA, and most of our DNA sequences, with chimpanzees and bonobos. The common ancestor of our species and of these two great apes walked the Earth up until 6 to 10 million years ago. If we accept an estimate of 3.9 billion years since LUCA lived, then humans, chimps and bonobos had identical lines of descent for 99.8 per cent of the time that life has been using the genetic code. For 90 per cent of those 3.9 billion years, humans and fish had identical lines of descent, and for approximately 75 per cent of them we shared a direct line of descent with bananas, rice and tulips. For between 15 per cent and 20 per cent of those 3.9 billion years we had the same line of descent as bacteria like *E. coli*. We share progressively less of our genetic sequences the further back in time we have to go to find a common ancestor with another species. Our DNA is consequently more similar to that of species of fish than it is to species of plants. What the tree of life reveals is that life has been diversifying and proliferating for a very long time. You are descended not only from apes but also from fish, jellyfish-like beasts, archaea like strain MK-D1 and bacteria.

The tree of life that biologists have drawn is remarkable. I can spend hours looking up species that I was unaware existed, and that in many cases no one has studied. As one moves forward in time from LUCA we first encounter bacteria, which then threw off a new branch, the archaea, which in turn split into two to produce the eukaryotes on a new branch. There are then endless splits that encompass yeast, plants and animals, until eventually we find *Homo sapiens* at the end of one of the branches. Sitting next to us on neighbouring ends of branches are chimps and bonobos.

It seems unlikely biologists will discover a new domain of life, so I expect these three major groupings to remain, but scientists are continually discovering new extinct and extant species in each of the three domains and will continue to do so as we catalogue more of life's remarkable diversity. Despite being unfinished, at the time of writing, the tree of life contained over 2.2 million species with nearly all of these organisms consisting of only a single cell. We are unsure quite how many species there are still to be discovered, but as we find them, they will fit somewhere into the tree of life.

Each species alive today differs from others not only in their genetic sequences but also in what they can do. Strain MK-D1 is able to live off the waste of hydrogen-gobbling bacteria, but I cannot. In contrast, I can just about run a mile, while strain MK-D1 cannot. I have a very different body shape to Woofler, or the bird singing in the silver birch tree outside, and these body shapes map to things we can do. I can hold conversations, climb a tree and peel a banana. Despite my attempts to stop him, Woofler can effectively chase down pigeons and squirrels and expertly dispatch them in a way I cannot. Neither Woofler nor I can suck nectar from flowers and we would be unable to repeatedly migrate 3,000 miles from Alaska to Mexico and back each year, a feat the 3.5-gram rufous hummingbird achieves in each of its three to five years of life. Each species is different, and each is adapted to a particular way of life.

Woofler can effectively hunt small prey, I can peel a banana and hold a conversation, and a rufous hummingbird can fly the length of the North American continent because we have attributes that enable us to do these things. I have dextrous hands, a voice box and a large brain; Woofler has sharp teeth, keen eyesight and a very strong hunting instinct; while rufous hummingbirds have long bills that allow them to feed on high-energy nectar deep inside flowers and wings that let them hover and migrate. Each species has characteristics that allow it to do a specific set of tasks that enable it to find food, survive and reproduce. Biologists call these characteristics phenotypic traits, and they are defined as the attributes of an organism you can measure. Phenotypic traits range from specific molecules that individuals produce through to an individual's body weight or aspects of personality.

Phenotypic traits differ between species. Dogs look different to people because they have a different set of phenotypic traits. Different individuals within a species can have different values of the same phenotypic trait. For example, you may be taller than me, or shyer, or have different-coloured eyes. Differences between species in the phenotypic traits expressed, and differences between individuals within a species, are the result of genetic differences. If we have different blood types from one another it is because we have different alleles at one or more of our genes. When individuals with different phenotypic trait values have different chances of surviving and reproducing because they differ in their ability to detect, acquire and utilise resources such as food, natural or sexual selection occurs, and this can result in different generations having different frequencies of specific alleles. Such genetic change across generations is evolution.

An individual's DNA sequence contains the code to produce proteins that determine how an organism develops, and the phenotypic traits it expresses. Phenotypic traits are important because they determine what an individual of a particular species can do, while the DNA sequence

Woofler's Taxonomy

DOMAIN	Eukarya
KINGDOM	Animalia
PHYLUM	Chordata
CLASS	Mammalia
ORDER	Carnivora
FAMILY	Canidae
GENUS	*Canis*
SPECIES	*Canis familiaris*

determines which phenotypic traits an individual of particular species will develop. The study of phenotypic traits has been the focus of much of my career post-Ph.D. Given my graduate studies at Imperial College London had focused on how animals impact the distribution of trees by dispersing and consuming seeds and seedlings, it might seem surprising that I switched tack and started studying phenotypic traits and evolution as I got older. I am glad I did, but the route I took was not carefully planned.

As I neared the end of my Ph.D., the realization that I would need to get a job dawned, and so I started to think about next steps. I knew I loved science and wanted to understand why I existed, but I didn't know the best way to achieve this. Was it to become an academic, working on one small part of the problem while reading up about other aspects, or should I take another job that might give me flexibility to do the reading I needed to do? I didn't really know. The first job I applied for was a postdoctoral researcher position at the Institute of Zoology, the research arm of London Zoo, where a postdoc is the next stage in academia after completing a Ph.D. Postdocs are typically employed on contracts that last for between one and five years with the job description outlining a particular scientific challenge that should be addressed. They are usually advertised by a university faculty member who has succeeded in securing funding from a government grant or a charity following a highly competitive peer-review process. As well as this job, I also sent off for forms to become a weatherman at the BBC, and for the civil service fast track, where postgraduates are trained to become scientific advisers within government departments. The detailed description for the weatherman job revealed I would need to be more interested in the weather than I was, and the civil service fast-track application procedure was so long and drawn out I could never motivate myself to start filling out the form, so the only job I ended up applying for was the postdoc position at the Institute of Zoology. I was amazed when I was offered it, as the interview was disastrous.

On hearing I had an interview, two of my housemates, Ana and Jane, insisted they take me shopping for a new suit. It was a sensible call, as my clothes no longer fitted me. I wasn't particularly domestic back then, and to get the washing machine going I simply turned the dial to the first setting, which turned out to be boil wash. I only learned of my mistake when Ana asked who was boil-washing their clothes, and I suddenly understood why they had become so tight. Under the sartorial gaze of Ana and Jane I bought a dapper suit, collared shirt and tie, and the next morning I headed off to my interview. I had to get from Sunningdale in Berkshire to north London in the rush hour, and this meant standing for most of the journey. As the train drew to a stop at Twickenham, the man standing next to me managed to spill most of his cup of coffee down my shirt and tie. Not a great start. Shortly afterwards I realized my journey planning was cutting things very tight, and a tube delay meant I had to run the mile from Camden Tube station to the Institute of Zoology on one of the hottest mornings of the year. I arrived drenched in sweat, coffee stained and smelling like a cross between Starbucks and a gym.

My interviewers, Steve Albon and Josephine Pemberton, gave me a few minutes to catch my breath before ushering me in. I was flustered and still sweating profusely, so much so that one of them offered me a towel to wipe my dripping forehead just after I'd answered the first couple of questions. Despite this, I must have impressed, for the following day

I was offered the job. I had a three-year position at the Institute of Zoology to study the genetics and ecology of red deer living on the Isle of Rum in the Inner Hebrides and a breed of wild sheep living on Hirta in the St Kilda Archipelago, one of Britain's remotest islands. Little did I know it back then, but I would continue to work on these two study systems for the next eighteen years. I became good friends with Steve and Josephine and many others who collaborated on the project, and I only stepped back from this study when a scientific disagreement over approaches threatened to jeopardize these friendships. I am pleased to say we all remained friends, and the move to work on other systems, including wolves in Yellowstone, guppies in the freshwater streams of Trinidad, and a small species of green bird called silvereyes that live on the islands off Australia, a study system my wife Sonya has worked on for over twenty-five years, proved to be a great decision.

In the years I spent working with red deer on Rum and Soay sheep on Hirta I became proficient in working with mark-recapture data, something I still do. To generate these data, individual animals living in the wild are humanely captured, often shortly after birth, and affixed with a unique mark. In the case of sheep, the mark is a numbered ear tag, for red deer it was an expandable patterned collar, in wolves it is a collar that sends out a bespoke radio signal that allows us to identify where the animal is, and in the guppies it is a tiny tattoo. After marking, animals are then released back into the wild at the same location they were captured, and each time they are subsequently seen or caught their location is recorded. Coupled with other data that are collected at the time of capture, including blood samples that allow family trees to be constructed for each individual, and phenotypic trait data on attributes such as body size, position in a dominance hierarchy, evidence of parasites, state of the animal's dentition, and all sorts of other measurements, it is possible to study questions as diverse as the effects of climate change on wild animal populations, and why and how they evolved the phenotypic traits they have.

The reason I was so happy with my decision to work with guppies, wolves and silvereyes I quickly realized was that my knowledge of the natural world was very biased by what sheep and deer do. As I started to work with experts who had dedicated their lives to building incredibly valuable, individual-based studies of other species, I came to appreciate that each individual in each species really is unique, not just in its genetic code but also in how it lives its life. I would never have understood the way that nature works in as much detail as I do had I remained a sheep and deer expert. Those studies were invaluable for my training, and I will always be grateful for the opportunity to spend so long working on these systems. But it was the guppies, the wolves and the silvereyes that allowed me to flourish. Hours spent chatting with research collaborators in Yellowstone National Park, and with Sonya on walks with Woofler in and around Oxford or on visits to Australia, helped me break free from my narrow herbivore focus.

What I started to appreciate from working with each of these very different species is that individuals in each population experienced very different causes of death, and these causes mattered. Wolves tended to die in fights with other wolves or from disease; guppies that lived with predators would inevitably end up in the jaws of a predatory fish such as a goby, while those that lived away from predators would die from a lack of food; while Sonya's silvereyes would often die during harsh winter storms, from disease outbreaks, occasionally from fights with other birds, or from unknown causes in old age. Just as important as the causes of death were events that prevented some individuals from breeding. In some cases, these events were a failure to lay down enough fat to produce offspring, or a failure to find a mate or breeding territory, or dominant individuals killing the young of subordinates, as happens in the wolves. Within each population, the major causes of death and failure to breed favoured phenotypic traits that reduced these threats, and by understanding what caused death and reproductive failure, and how different traits altered these risks, I could draw generalizations across very disparate study systems by linking their evolution and ecology. Natural selection is all about understanding how different phenotypic traits help individuals avoid things that can kill, while sexual selection focuses on how other phenotypic traits minimize the chances of failing to breed. Individuals with DNA sequences (genes) that code for the development of phenotypic traits that reduce the risks of death and reproductive failure leave more descendants. More descendants mean more copies of genes that code for the production of these beneficial phenotypic traits, and this means these beneficial genes increase in frequency within the population from one generation to the next. Evolution driven by this process has been continually ongoing since the first organism evolved.

Across the 3.9 billion years of the line of descent from LUCA to you and me, our ancestors have faced countless causes of death and things that could stop them reproducing and passing on their genes to the next generation. Yet each of our ancestors survived long enough to produce descendants, with evolution eventually producing us. The same is true for every other individual of every other species alive today. To achieve this astonishing feat, numerous challenges were faced and overcome, with evolution producing ingenious solutions to minimize the myriad risks our forebears faced. As the tree of life grew and life diversified, evolution moulded endless new phenotypic traits to help each species cope with the causes of death and reproductive failures that life threw at it. These phenotypic traits are coded in DNA, with the study of life's most important acid having helped biologists piece together not just the tree of life but also much of its history.

For many genes, the two alleles are always identical in all individuals within a population, but for others they are not. For example, alleles at genes that control the development of our two arms and two legs are always the same within each of us, while alleles at genes that determine our hair colour can differ. It is genes that have more than one type of allele that make each of us different. The more closely related you are to someone, the more similar your genomes.

The further apart two species are on the tree of life, the more different their genomes are. Time has led to their genomes diverging since they shared a common ancestor. However, not all parts of DNA diverge at the same rate as we move backwards along our line of descent towards LUCA. Some genes have changed very little with time. For example, genes associated with translating the genetic code into proteins are said to be highly conserved, for they have hardly changed in the last 4 billion years. Other genes can differ substantially between even closely related species. Heterochromatin, for example, is an important gene that is involved in switching other genes off, and it differs in very closely related species of fruit fly.

New alleles, and new genes that produce new proteins, arise via mutations to the genetic code. Mutations occur because errors are made when strands of DNA are copied when cells divide. A mutation is a copying error, and it is how the Sars-Cov-2 virus mutated to create new strains.

Mutations appear to occur at random, with many impacting the way an organism develops. These impacts can be mild, resulting in outcomes such as my impaired vision due to a failure of my retina to develop properly. Other mutations can be much more debilitating, resulting in blindness, deafness, premature ageing, early death, chronic pain, Downs syndrome and some types of personality disorder. Deleterious mutations consequently prevent development proceeding as it should. However, a small proportion of mutations are advantageous. They result in phenotypic traits being expressed that allow individuals to better avoid death or reproductive failure. It is these advantageous mutations that have allowed life to thrive, to increase in complexity on some branches of the tree of life, and to conquer the world. Without these advantageous mutations, we could never have evolved from LUCA. However, the cost of life's success is the deleterious mutations. The cost of our existence is countless premature deaths, suffering and failures to thrive of endless organisms that had flawed self-assembly manuals, which meant they left no descendants. Mutation and selection drive evolution, but because mutation can happen anywhere in the genome it means some individuals lose life's developmental lottery while a few others win.

The simplest type of mutation occurs when a cytosine, thymine, adenine or guanine is not copied correctly. Such a mutation in one of my ancestors is responsible for my poor eyesight (more on that later). An example of a point mutation such as this might mean a cytosine nucleobase is accidentally copied as adenine at some point on chromosome 14. The new allele that is produced might change a single amino acid in the chain the gene codes for, and this could mean the protein it produces takes a different shape. The protein's changed structure might slightly alter the way a phenotypic trait develops by slowing down a chemical reaction compared to the non-mutated allele. If the new phenotypic trait gives individuals that carry it a survival or reproductive advantage, the newly mutated allele should start to spread within the population, but most mutations are deleterious and are quite quickly lost from the population.

Single-point mutations, where one nucleobase is switched to another, is just one type of mutation. Other types involve more substantial changes. Perhaps the most intriguing are mutations that arise where 'junk DNA' that didn't produce a protein starts to. For a sequence of DNA to start producing a chain of amino acids, the molecules in your cells that read the DNA sequence and translate the code into a protein need to encounter a start codon that initiates the production of a string of amino acids. There are a number of nucleobase triplets that initiate the production of an amino acid chain, but ATG is a common start codon. There are also codons that tell the machinery to stop. Every now and then a point mutation will produce a new start codon in a stretch of junk DNA that was not used to produce proteins. When that happens, a completely new protein can be produced. In most cases, it will not serve a useful function, but on rare occasions the protein may find a role for itself. Mutations such as these are uncommon, but they can produce completely new phenotypic traits that have never been seen before by life.

We tend to think of viruses as being nothing but trouble,

but they can also occasionally be the source of new, useful genes. Viruses, such as the coronaviruses that cause COVID or measles or rabies cannot reproduce on their own. They need to use the machinery of your cells, and they do this by infecting you. Once inside your cells, they hijack the cellular machinery to make copies of their genetic code and the proteins that coat their exterior, before self-assembling as new virus particles. Most cells that are infected by viruses die, but not always. The genetic code of some viruses can end up being permanently spliced into the DNA of an infected cell. If this happens in the stem cells that make sperm and eggs (the germline), then the new gene can be passed on to offspring, and can then be used to make proteins and new phenotypic traits. On most occasions these proteins may not have a useful role, but they sometimes do, and viral genes co-opted by hosts are known to have played a role in the evolution of the placenta in early mammals.

There are even more types of genetic mutation, including events known as deletions, insertions and inversions. Stretches of DNA can get copied from one part of the genome and inserted into another part, while DNA sequences can also get flipped around, appearing in reverse order. Sometimes two chromosomes can join to form a larger new one, while others may split in two. On occasion, the whole genome can be duplicated, doubling the number of genes by making a new copy of each one. Recent research has even revealed genes jumping from parasites into their hosts. DNA is dynamic and subject to change.

The history of mutations in our lines of descent means different species have different genes, and different alleles at some of the same genes, and they also have different genome sizes. The Japanese alpine plant *Paris japonica* has a genome consisting of 149 billion nucleobases. The human genome is fifty times smaller, consisting of a paltry 3 billion or so nucleobases. Compared to the bacteria *Nasuia deltocephalinicola* that lives inside insects, this is still vast. Its genome consists of only 112,000 nucleobases, nearly six orders of magnitude fewer than *P. japonica*. Some species appear to require more DNA than others to live their lives, although, perhaps surprisingly, the size of the genome doesn't necessarily correspond to the number of genes. The plant *Arabidopsis thaliana* has 27,416 genes packed into a genome of 135 million nucleobases, while Norway spruce trees have about the same number of genes but a genome of 19 billion nucleobases.

The way that DNA is organized also differs between species. The atlas blue butterfly is the animal with the largest number of chromosomes, with 450 arranged into 225 chromosome pairs. In contrast, females of the jack jumper ant have only one chromosome pair. Males of the species, as in all ant species, have only one copy of each chromosome rather than two. Biologists describe them as haploid rather than diploid as their chromosomes do not come in pairs. Remarkably some species of plant have three or more copies of each chromosome. The adder's tongue fern can have up to ten copies of each, while even the humble banana is triploid, with its chromosomes coming in three copies. Genome size, the number of chromosomes and the number of copies of each chromosome are phenotypic traits, and so can evolve much like any other.

By comparing genome similarity across species, biologists can piece together which species are closely related, and which are not. However, when an organism dies its DNA can quite quickly break down except for in very unusual circumstances. In the film *Jurassic Park*, dinosaur DNA is extracted from insects preserved in amber, but even when locked up in hardened tree sap, DNA cannot last anywhere near the 66 million years it would need to survive from the Cretaceous-ending meteor impact until now for us to genetically sequence a dinosaur. The oldest DNA yet found was beneath the Greenland icecap and is about 2 million years old. Because most DNA degrades quite quickly after death, we can only hypothesize about the size, structure and make-up of the genomes of *T. rex*, the first mammals and the common ancestor we shared with MK-D1 billions of years ago. We may be able to assert, reasonably confidently, that an extinct species had a particular gene if descendants from a shared ancestor with that species alive today all have the gene, but that only takes us so far.

DNA has allowed biologists to piece together the tree of life in astonishing detail, but it is of less use in working out what extinct species looked like. Even if we can assume an extinct species had a particular gene, that does not mean we know with certainty how the gene influenced the species' development or phenotypic traits. The same gene can often be co-opted for multiple different uses. For example, the protein GroEL is found in many bacteria species, and it plays a key role in ensuring that chains of amino acid fold into the correct protein shape. The same protein, produced by exactly the same gene, is also used as an insect toxin by antlions.

The larvae of a type of lacewing are known as antlions, or doodlebugs. These insect larvae dig cone-shaped pits in sand, partially burying themselves at the bottom. Unsuspecting insects, often ants but sometimes beetles, lose their footing on the edge of the sandpit and slide to its base, where the antlion injects them with a paralysing toxin before consuming their prey's innards and discarding the inedible exoskeleton.

The toxin is produced by bacteria that live in the antlion's salivary gland, and it is the same protein used by bacteria to fold proteins. Many, and perhaps the majority of genes, produce proteins that are put to more than one use, and that makes it hard to know how an extinct species might have used a particular gene or which phenotypic traits it helped produce.

Different species also switch on genes for different lengths of time during development. Fish species that grow proportionally larger fins turn on some genes for longer as they are growing their fins than species that grow smaller ones. Many differences in body shape and organ sizes between mice, monkeys, humans and other mammals are due to different species switching genes on and off at different points in development, rather than because they are using different genes. A gene associated with brain development consequently may not tell you much about how big a brain an extinct organism might have had. We need to turn to fossils to understand the form of life past.

People have been aware of fossils for at least two and a half thousand years. The ancient Greeks realized that fossilized shells were once living shellfish, and when Xenophanes of Colophon uncovered such fossils, he correctly surmised that the land on which he had found them must once have been under the sea. Although many cultures from lands as far afield as China and the Middle East were aware of fossils and appreciated that they were records of long-dead organisms, it was not until the 1840s, when Richard Owen, the founder of London's Natural History Museum, coined the name 'dinosaur' that interest in fossils became mainstream. In retrospect, it became obvious that humans had stumbled across dinosaur fossils for millennia, but had attributed them to mythical monsters, including giants. Richard Owen's breakthrough was to analyse bones from three different species of dinosaur and realize they all came from an extinct group of reptiles that could grow to enormous size. Dinosaur translates to 'terrible lizard', and once the public appreciated that monstrous beasts really did once walk the Earth, our fascination with dinosaurs began.

Dinosaur fossils are reasonably common, although the vast majority of dinosaur corpses rotted away leaving no trace. Their fossils are common because dinosaurs were incredibly successful, roaming the planet for 165 million years, and some species would have had large populations. Big dinosaurs also had large bones and teeth, and it is the hard parts of animals and plants that are most likely to fossilize. In comparison, individual cells or soft tissues such as muscles, brains and flowers rarely form fossils. Nonetheless, in some cases softer tissues have been preserved. Despite this, there has been a lot of life on our planet over the last 4 billion years that we know little about, even though palaeontologists have got very good at predicting where best to look for fossils from each past geological epoch.

It is not always straightforward to tell whether an animal is an adult of a small species or a juvenile of a larger one. As species develop, their body shape can change radically, and when only a few bones or teeth are fossilized, interpreting them can be hard. Juveniles and adults of a species can differ substantially in both form and size. Emu chicks are hatched at about 5 inches tall, about the size of the world's smallest flightless bird, and are fluffy and remarkably cute. Yet they grow to a height of five feet seven, with the adults being monstrous and bearing little resemblance to the young.

A few years ago, while visiting my father-in-law's farm in Queensland, we were brought a day-old emu chick that had become separated from its father. Emus, and other ratites such as ostriches, are unusual for birds in that it is the father that does all the parental care of the chicks. Young birds of most species are not as attractive as the adults, being born without feathers. Emus, chickens and ducks are different, in that their young do have feathers, and the emu chick we were given to care for was cute. And if we didn't look after it, it would die.

I originally named the emu Emma, but on pondering whether it might in fact be male decided we needed a gender-neutral name. Emma-Steve imprinted on my father-in-law, and he on it, and he raised it on his farm. For months the bird was a beloved pet, and it turned out it would eat pretty much anything, including the keys to the tractor, but it was most partial to steak. Most emus eat a diet less protein-rich than Emma-Steve's, but given the cattle on the farm, steak was not in short supply. By the time Emma-Steve became an irascible 'teenager' at about a year of age, it was the biggest emu we had ever seen, growing to about six feet tall. If we hadn't known what emu chicks looked like, we would have been in for a nasty surprise.

It turned out that Emma-Steve was a him, and he developed from an extraordinarily cute chick into a violent beast with a strong desire to top the dominance hierarchy. After Emma-Steve chased my daughter Sophie into the house, attacked my father-in-law and generally deterred all visitors, it was decided it was time to take action. He under-went what biologists refer to as a soft release, which meant we left the gate open before shutting it once he'd left. But we would regularly see him around. His large size did him no harm, and on one sighting we were excited to see him with a clutch of eight young.

My father-in-law has raised all sorts of orphaned animals on his farm, from wallabies to stray puppies. He has vowed to never again raise an emu. Emma-Steve's initial cuteness deceived us, and even though we knew what sort of monster the chick would grow to be, we seemed to forget that in our desire to save the young chick's life. Juveniles and adults of a species can be very different, but despite the challenge that development poses to palaeontologists, their careful study of the fossil record, coupled with the insights that geneticists studying DNA have produced, means we have a good history of life on Earth.

LUCA was a simple bacterium that used the same genetic code that we rely on but was otherwise very different from us. Attempts to work out what its genome looked like suggest that it probably ran its metabolism on hydrogen sulfide produced by volcanoes rather than by eating plants, fungi and animals. Oxygen would have killed LUCA, while we die without it. Its genome would have been tiny compared to ours, and sex was alien to it, while it is necessary for us to reproduce. There were very many mutations, some of which led to novel phenotypic traits, which helped our ancestors survive and reproduce, that were required for you and me to evolve from LUCA, and these played out over nearly 4 billion years. The construction of the tree of life through the study of DNA and fossils, coupled with key events in the history of our planet such as meteor impacts and periods of extreme volcanic activity, has allowed biologists to work out what had to happen for LUCA to evolve into humans. There is still much to learn, but we know a lot. The process of mutation impacting development with natural and sexual selection operating on the resultant phenotypic traits led to us. The molecular mechanisms involved are often very complicated and are beyond the scope of this book, but I will briefly consider a few of the most important evolutionary innovations on the journey from LUCA to us.

As life secured its footing on our planet it became adept at using a wider range of energy sources than the hydrogen sulfide that LUCA likely relied upon. By at least 2.7 billion years ago, and perhaps as long as 3.5 billion years past, life had found a way to run its metabolism using light. The first bacteria to do this were cyanobacteria, the type of bacteria that produce the stromatolites found at Shark Bay in Western Australia and in the Bahamas. They use photosynthesis, something that all plants, algae and many bacteria rely upon today.

Photosynthesis uses the energy from photons to break apart molecules of carbon dioxide and water before using the freed carbon and hydrogen atoms to build the amino acids needed to make proteins. In doing this, oxygen is produced as a waste product. Although each individual cyanobacterium produced very little oxygen, with billions of them releasing it into the environment over millions of years, all the elements in the Earth's crust and atmosphere that could react with oxygen eventually became oxidized, at which point oxygen concentrations began to increase in the atmosphere. Unreacted oxygen first appeared in the Earth's atmosphere 2.33 billion years ago, steadily increasing in abundance over the next few million years until it constituted 3–4 per cent of the atmosphere. This is only a fraction of 21 per cent of the atmosphere that oxygen constitutes today.

The cyanobacteria were the first global polluters, and although life went on to use the oxygen they produced to great effect it was poisonous to most forms of early life. Billions of bacteria were killed by cyanobacteria's oxygen. But one species' poison is another's meat, and life found a way of using oxygen. The endless puffs of oxygen from countless cyanobacteria billions of years ago paved the way for you and me to exist. The great oxygenation event in which oxygen increased to 3–4 per cent of the atmosphere was a necessity for our evolution, but it did not foretell it.

An increase in oxygen in the atmosphere allowed a new type of metabolism to evolve, with organisms adapting to use the reactive nature of oxygen to gain energy from glucose, a type of sugar that is used by a very wide variety of life forms. As evolution has shown time and again, when a new resource becomes sufficiently abundant, life finds a way to use it. Once life had mastered how to use oxygen to break down glucose it changed the world beyond recognition, allowing new ways of life and organisms built from more than one cell to evolve.

All animals alive today run their metabolisms off glucose sourced by breaking down large molecules called carbohydrates. Life uses carbohydrates to store energy that can be used at a later date, and for building other large molecules. Every cell contains carbohydrates, so the ability to run metabolism and DNA replication using glucose as a fuel opened up the opportunity for life to thrive by feeding off the living or recently dead. These lifestyles began with single-celled organisms feeding on one another, eventually via endocytosis, but led to predation, herbivory, parasitism and scavenging. Oxygen was consequently a double-edged sword. Not only was it a poison for many bacteria, it also heralded the arrival of a more dangerous world, as the world now contained a new way to die: death by other organisms.

As oxygen levels increased, cells evolved large sizes as more oxygen could diffuse deeper into them. These larger cells required more organizational structure than bacteria and archaea to run efficiently, and internal membranes evolved to help direct and shuttle chemicals around within the cell. These early eukaryote cells could be thousands of times larger than bacteria. Research into the evolution of eukaryotes is a highly active field, with many hypotheses proposed as to how the structures in eukaryote cells evolved. Currently we are not entirely sure how archaea evolved into eukaryotes, but there is one key event worth mentioning. At some point shortly after eukaryotes emerged approximately one and a half billion years ago, a failed predation event occurred when an early eukaryote cell attempted to consume a bacterium. The bacteria survived and it ended up setting up home inside the cell, where it divided to produce offspring. Over time, the early eukaryote and the bacteria evolved a highly effective symbiosis where the early eukaryote provided safety and glucose to the bacteria in return for it running the cell's metabolism. The bacteria generated the adenine triphosphate, or ATP, that is the end-product of metabolism and is the fuel used to run the chemical machinery in all cells.

Descendants of these bacteria are found in all eukaryote cells, and they are called mitochondria: organelles with their own membranes and DNA. Mitochondria can no longer survive outside eukaryote cells, and eukaryote cells die without them. When our mitochondria cease to function, we die. Something else normally kills us first, but a key reason we age is because our mitochondria wear out. Sometimes the DNA in mitochondria experiences a mutation, meaning they do not work appropriately. When that happens it usually results in an early death. All eukaryotes, including plants and fungi, have mitochondria, with plant cells hosting the descendant of another symbiotic prokaryotic in their cells. These organelles are called chloroplasts, and this is where photosynthesis happens, with light being used to produce the glucose used to run plant metabolism. All complex life on Earth today owes its existence to mutualisms between bacteria and early eukaryote cells that were closely related to strain MK-D1.

Some eukaryote species, including humans, have sex to reproduce. Sex evolved quite early in eukaryote evolution in single-celled organisms similar to yeast. However, not all eukaryotes reproduce by sexual reproduction. In some species, females produce genetic clones of themselves, making males redundant, and this poses a paradox. Every time a clonal female divides, she produces an exact genetic copy of herself. Furthermore, every daughter can produce genetic clones herself. In contrast, every time a sexually reproducing female breeds, on average, only half of her offspring will be female and able to give birth to new descendants; the other half are male and they cannot give birth to new descendants. Each offspring born to a sexually reproducing female also has only 50 per cent of her genome, represented by one chromosome in each chromosome pair. All else being equal, genes in the asexually reproducing clonal strategy would spread through a population at twice the speed of the sexually reproducing strategy. Sexual reproduction would be outcompeted by asexual reproduction, and the sexually reproducing strategy would be driven extinct. Sex is consequently evolutionarily costly, so for it to evolve, there must be some advantages.

Biologists have identified a number of advantages of sex. For example, sexual reproduction allows for evolution to more rapidly rearrange the way a genome is organized, helping populations rapidly adapt to evolving viruses and bacteria. Sexual reproduction can also help populations rid themselves of deleterious mutations. Benefits such as these must have outweighed the costs of sex early in eukaryote evolution, and although sexual reproduction is not ubiquitous, it is the way that most animals and plants reproduce. Having evolved in single-celled eukaryotes, sex was adopted by more complex species once multicellularity evolved.

At some point between 600 million and 1.6 billion years ago, multicellular life appeared. Most life forms, including all bacteria and archaea, consist of a single cell that can survive and reproduce without relying on other cells. Multicellular life such as fungi, plants and animals is made up of lots of cells working together, with none of these cells able to thrive alone. In such organisms, each cell in the body has the same genome, but the cells that develop from it can be classified into different types, with each type performing different functions. For example, your skin cells are different from your nerve cells, which in turn are different from muscle cells, but they all contain the same DNA sequence wound up within a membrane called the cell nucleus.

Evolving multicellularity required two problems to be solved. An organism consisting of multiple cells had to develop from a single initial cell and, once the organism had matured, it needed to produce new single cells that could develop into new individuals. In sexual reproduction, initial new cells are the result of a fertilized egg. Most animals do this by making sperm and egg cells that have only one copy of each chromosome, making them haploid cells. Conception involves one egg and one sperm cell fusing to make a new diploid cell that then makes a copy of itself, before each of these two cells themselves divide to make four cells. The process is repeated until a ball of sixty-four or so cells called the blastula forms. The blastula starts to differentiate into different cell types, with the first differentiation into cells that become the placenta and those that become the embryo. Whether one of the blastula cells develops into a placenta or an embryo cell is determined by the concentration of chemicals produced by neighbouring cells. Those on the outside have fewer neighbours, so experience lower concentrations of some chemicals, and this lower concentration results in some genes being turned on while others remain off. In contrast, higher concentrations of these chemicals in cells deeper in the developing cell ball mean a different set of genes are turned on and off and the cell develops into another cell type.

Once the first differentiation has occurred, the two different types of cells produce chemicals that are often unique to that cell type. These molecules leave cells, and form chemical gradients, such that there are more molecules closer to the source cell type than there are further away. Concentration gradients of many different chemicals throughout the ball of developing cells provide the instructions to each cell as to which genes to switch on and which to turn off, and these determine the type of cell it becomes, and more and more different cell types are produced. Throughout development these chemical signals lead to muscles, kidneys, eyes, ears, heart, fingernails and all the different bits that you are up of arise from the same genome. In remarkable demonstrations of how important variations in chemical concentrations are for development, biologists have manipulated chemical gradients in fruit flies and other laboratory animals by turning specific genes on and off at various ages to produce monstrous flies.

You consist of about 30 trillion cells, classified into about

220 broad types. There is further variation within each type, and some scientists argue we should classify at a finer scale. During development these cell types organize themselves into the different organs and tissues that constitute you. The same genome produces these very different cell types by turning on and off different genes during development and by leaving some genes functioning for longer in some cells than in others. Turn one set of genes on to make proteins and you end up with a cone cell used for vision, turn on another set and you create a fat cell for storing energy. The self-assembly manual that is your genome contains instructions that determine when to turn particular genes on and off during development, creating concentration gradients of molecules that instruct other genes in other cells to be turned on or off.

Animals are the champions of cell differentiation, but plants, fungi and some types of algae such as seaweeds do it too. These other groups produce only a handful of different cell types from their DNA and have far fewer organs than animals. Construction of the tree of life reveals that multi-cellularity in eukaryotes evolved independently at least ten times. But how did it arise?

Over the billions of years that life was restricted to single-celled organisms, mutation and selection produced an ever-increasing number of useful genes. Bacteria evolved various ingenious ways of sharing DNA, and by this process some single-celled species became masters of multiple ways of life. Different genes would be activated depending upon the environment in which the cell found itself. For example, if oxygen was available to power metabolism, then glucose could be used as a source of energy. In contrast, if it was absent, try another source of carbon atoms. Flexibility to use different genes in different environments doubtless paved the way for the evolution of multicellularity, but it was only a single step of the many that were required.

A key challenge in making the first multicellular organism would have been to clump cells together. The most likely way for this to have happened was for dividing cells to fail to completely separate, such that a ball of genetically identical cells formed. However, that is not the only route to producing a ball of cells. Another hypothesis is that groups of cells from different genetic lineages started to cooperate and, via a mechanism that is yet to be identified, merged their DNA into a single genome. Genetically different cells of some simple species are observed to group together today, so there is some support for part of this hypothesis, but until evidence of genomes merging is documented, this will remain an unproven route. A third idea is that a single cell produced multiple nuclei, each containing a copy of the genome, before membranes separated the nuclei and nearby organelles, producing a group of cells. Such behaviour has been observed in some modern-day species, but it is relatively uncommon. The fossil record does not record processes such as those that resulted in multicellularity, so we may never know for sure how multicellularity arose, but it is possible that across the ten or more independent origins of it, evolution used each of these three methods.

The first multicellular animals likely consisted of a ball of cells with little differentiation between them. Nonetheless, natural selection favoured these simple balls. The first multicellular organisms may have been able to consume single-celled competitors. They may also have been protected from large single-celled predators, and they might have had a mobility advantage. In photosynthetic species, multicellular species could have been superior competitors for light, growing over the top of single-celled species. But there would have been costs too. Resources such as sugars and oxygen would need to be transferred to cells in the interior of the organism, for example, but the benefits outweighed the costs, and multicellularity was here to stay.

Multicellularity does carry several costs. For example, multicellular organisms delay reproduction while they develop, and this means that single-celled organisms can make copies of themselves at a faster rate than multicellular ones. All things being equal, because single-celled species can reproduce more quickly than complex animals, they should outcompete their larger competitors. Multicellular species also inevitably die. Although not an evolutionary cost, it is an undesirable feature of being a complicated organism whose bodies are divided into germ cells and the disposable soma. If the germline has done its job effectively it will be represented in offspring. All of your body, from your limbs to your brain, your eyes to your toes, is disposable soma, existing solely to maximize the probability you will produce offspring, each of which will carry half your genes.

In animals, plants and fungi the benefits of multicellularity must have outweighed the costs. However, in other eukaryotes, such as yeast and amoebae, it was not favoured. Most life on Earth is still unicellular. It is a very effective way of life that has persisted for billions of years longer than multicellular life has been around. We tend to think of multicellular life as evolutionarily superior, but that is not correct. It is just one way that life found to pass genes from one generation to the next. But how did a simple ball of cells evolve into complex animals like you and me?

Fossils reveal that in one multicellular lineage, the simple balls of cells evolved into tiny bowl-shaped animals. The bowl may have been a primitive trap for single-celled organisms that were then broken down by chemicals these primitive animals produced. Different forms evolved in the ancestors of plants that created branching structures, maximizing the surface area that could use light to photosynthesize. Multicellular life started to diversify, much as unicellular life had done, using any available resources it could find. A key step had been taken on the line of descent from LUCA to us.

Over the next half a billion years, more and more complex species evolved. There was a spectacular burst in the evolution of complex animal life lasting 13–25 million years just over 500 million years ago. The Cambrian explosion, named after the geological period that began 539 million years ago and lasted for over 53 million years, was when our ancestors, and indeed ancestors of all animals with a backbone, first appeared in the fossil record. High-tech imaging of sixteen fossils of a species called *Pikaia gracilens* revealed it had a nerve chord that, over the next few million years, developed into the backbone found in all vertebrates. Like many distant relatives, *Pikaia gracilens* would not have been an inspiring dinner-party guest, although it certainly would have stood out. It was about five centimetres long and looked like a primitive eel with tentacles on its head. There is a related group of animals alive today called lancelets, which resemble *Pikaia gracilens*. They live in shallow seas from the tropics to as far north as Scandinavia, and they feed by filtering bacteria, algae and small animals out of the water column. *Pikaia gracilens* pioneered this way of life that has survived for over half a billion years.

It is not just our ancestors that first appear in the fossil record during the Cambrian explosion. Some of the forms that evolved then have gone on to evolve into mussels, sponges, octopuses and lobsters. Evolution had mastered multicellularity, cellular differentiation and the use of chemical gradients, and during this period evolution produced not only the ancestors of all animals alive today, it also experimented with body forms that failed to survive evolution's filter. The one species I wish had survived was *Tamisiocaris borealis*, which measured nearly a metre in length. It was closely related to an apex predator called *Anomalocaris*, an ancestor of modern-day shrimps and crabs, but *Tamisiocaris borealis* was altogether a much less fearsome beast. It would not have looked out of place in the seas of Pandora in James Cameron's *Avatar: The Way of Water*. It fed on bacteria and algae using delicate bristles on its front arms, or legs, or whatever we should call them, to entrap them. If it was still alive today, it would be a marvellous addition to any large marine aquarium.

The Cambrian explosion was also the geological period when the modern-day animal head evolved. Evolution had produced cells and eventually organs that could sense light, electric fields, touch and sound waves, and likely taste and smell. Primitive brains evolved alongside these early sense organs, and during the Cambrian explosion eyes, ears and noses migrated to the same part of the organism that became a head and contained a brain. Although there were peculiar-looking beasts such as *T. borealis*, many animals that evolved during the Cambrian explosion would not look entirely out of place in today's seas, as it was the period when many features of modern-day animals first appeared. But it was to be nearly another half a billion years before individuals of a descendant species of *Pikaia gracilens* gazed upon its fossils with wonder.

Towards the end of the Cambrian explosion, when animals rapidly diversified in the oceans, seeding the way for vertebrates, plants began to colonize the land, with the earliest evidence suggesting that the first species took root (literally) about half a billion years ago. Bacteria had been thriving in freshwater lakes and shallow seas for many hundreds of millions of years, and single-celled algae would have also colonized the same habitats once they evolved, but the first multicellular land plants came later. The earliest of these species would have grown roots in the sediment of lakes but started to evolve structures that allowed them to grow above the water to compete for light, in a similar way that mangrove species grow today. These early plants next adapted to thrive in seasonal lakes before colonizing areas that were flooded less frequently. As plants spread, organic soils followed, with earth and mud becoming commonplace about 400 million years ago.

Insects evolved a little under 500 million years ago, shortly after the first land plants established, and were the first land animals. The earliest insects looked a little like modern-day silverfish, and were yet to evolve flight. The first fossils of flying insects are 400 million years old, with flight evolving as a way of dispersing to new areas and of avoiding predation from the vertebrates that were by then well-established on land, having left the oceans 450 million years ago. The land plants and insects offered a resource that could not be resisted. A few species of fish had already evolved primitive lungs to gulp air as a way of supplementing the oxygen they extracted from the poorly oxygenated waters in which they lived. These fish may have also used land as a refuge from predators, or even moved short distances over land between temporary lakes or salty rock pools. Some modern-day fish still behave like this. Hart's killifish jump out of water to escape predators, and they move between streams across the forest floor in tropical South America. Phenotypic traits such as these would have given our primitive fishy ancestors a head start in colonizing land. Over time, limbs evolved from fins, and gills were eventually lost altogether.

As plants thrived on land, they began to compete for light, and being tall was advantageous. To grow upright to large heights, land plants evolved a rigid protein called lignin. They could use this to defy gravity and grow away from the ground below. Lignin is quite a tough molecule. When plants die today, and trees fall in the forest, lignin and other parts of the plant are decomposed by bacteria and fungi, but it can take years. By 300 million years ago plants were using lignin to great effect, and vast forests covered the land. The climate was warm and damp, and oxygen levels were higher than today, with dragonflies and other insects up to a metre in length flying between the trees. The ground beneath the trees was often swampy, and when plants died they were not broken down by bacteria or fungi. Some scientists have argued that this was because fungi and bacteria were yet to evolve enzymes to break lignin down, but this hypothesis is contentious. However, even if bacteria and fungi had evolved these enzymes, they were unable to deploy them in the damp environment of the forest floor, for in those swamps dead plants did not get broken down but instead built up. Over time they became compacted by yet more dead plants and then by sand and soil, and over millions of years the dead organic material became coal. We run our power stations, cars and aeroplanes by burning the remains of ancient plants that past life had found no way to exploit. Animals are yet to evolve a way to digest lignin, but humans have found a way to exploit coal (and oil, which formed from ancient algae and bacteria in shallow seas) and in so doing we are changing our planet. Carbon dioxide levels in the atmosphere have increased to levels not seen on Earth for millions of years, and the energy from long-dead trees is used to power machines that cut down living ones.

Although our addiction to coal and oil is changing the world and is a contributing factor to the very large number of extinctions that have happened in the last 300 years, without the coal and oil that formed in the Carboniferous geological epoch, modern-day civilization would not exist. We would not have built up the vast amount of knowledge we have of the universe, and nor would we have invented computers, cars and electric ovens. Coal and oil are both a blessing and a curse.

The mass extinction humanity is causing is referred to as the sixth great extinction our planet has experienced, although there may have been more. The most famous happened 66 million years ago when a meteor killed off the dinosaurs, but other mass extinctions we know of occurred 440, 360, 250 and 210 million years ago. Scientists can read the devastating impact of these events from the fossil record, and each time a significant fraction of life on Earth died. The biggest mass extinction was the one that occurred 250 million years ago, marking the end of the Permian geological era. It is commonly referred to as 'the great dying'. It wiped out over 80 per cent of species in the ocean, and 70 per cent of land-living vertebrates. It was caused by the huge amount of carbon dioxide released by the formation of the Siberian traps we encountered previously.

Mass extinctions often wipe out dominant species, opening the way for a new group to thrive, and the great dying was no exception. At the end of the Permian, our ancestors, a group called the synapsids, were the dominant species on the planet. The ancestors of dinosaurs, called sauropsids, were both rarer and smaller. The great dying killed off many though not all of the synapsids, including the two-ton herbivore *Tapinocephalus*, the largest animal of its time. It also killed several species of sauropsid, but those that survived got the upper hand in the new world, dominating it for the next 185 million years as dinosaurs, while relegating our synapsid ancestors to a life on the margins.

The synapsids had been the dominant group on Earth for over 100 million years, and, like all life forms, they evolved from an ancestor. The mass extinction prior to the great dying, which happened 110 million years earlier, had given a group of vertebrates called the amniotes a shot at being the dominant vertebrate fauna. Amniotes had evolved from fish and amphibian ancestors, with the first appearing 330 million years ago. They had the advantage of being less dependent on water than their ancestors, which helped them thrive in warm, dry climates. Their name comes from their ability to produce embryos surrounded by an amnion or amniotic sack, a membrane that prevents the developing foetus from drying out. Reptiles, birds and mammals are all amniotes, even though some lay eggs while others give birth to live young. A few million years after the first amniotes evolved, their branch of the tree of life split into the synapsids that were to evolve into mammals, some non-mammalian synapsids that subsequently died out, and the sauropsids that included the ancestors of dinosaurs, reptiles and birds.

Some of the extinct synapsids looked a little like modern-day carnivores, and they shared many traits with modern-day mammals. They were warm-blooded and had primitive mammary glands for provisioning their young, even though they laid eggs. Their jaws contained molars, canines and incisors, and some species appear to have had fur. The big synapsids including *Tapinocephalus* are not seen after the great dying, but some smaller species survived it, and by 160 million years ago had evolved into the first mammals. These mammals did not immediately go on to evolve into modern-day mammals like rhinos, hippos and humans because the surviving sauropsids, for reasons we do not understand, stole a march on the surviving synapsids following the great dying, growing to become the dominant class.

The dinosaurs that evolved from the sauropsids that survived the great dying roamed the Earth from about 240 million years ago. Like their famous descendant, *T. rex*, the first dinosaurs walked upright on their two back legs. They quickly rose to dominance and, once there, they enjoyed that position until 66 million years ago. Many dinosaurs were successful large species, and their success prevented mammals evolving to the sizes they achieve today. Mammals couldn't effectively compete with the tyrannosaurs, sauropods, triceratops and other large species. Interestingly, there were no small dinosaurs. It is possible that the small mammals that survived prevented dinosaurs evolving small sizes, while the large dinosaurs prevented mammals evolving to large sizes.

There are relatively few ways of making a living if you are a vertebrate. You get your food by feeding on other forms of life. Species can feed on vegetation, on seeds or fruits, on other animals they have hunted, or those they find dead, and, in the case of vampire bats and occasionally the vampire ground finch of the Galapagos, on the blood of living animals. No vertebrates are able to survive by consuming decaying plant matter, a way of life termed detritivory, but many vertebrates are omnivores eating a varied diet. Grizzly bears, for example, eat fruit, vegetation, insects, fish and meat. Regardless of what their favoured food is, evolution has produced a plethora of ways for species to detect, acquire and then utilize food.

The first step in finding food is to detect it. Animals have senses to help them do this, and these range from vision, smell, and touch, which are very familiar to us, through to echolocation, used by some bats. Each sense requires a specific organ – eyes for vision, ears for sound, a nose for smell – plus an area of the brain to make sense of the information the sense organs collect. Some food types are easier to detect than others, with it being easier for a bison to find grass than it is for a woodpecker to find bugs in bark. When resources are hard to find, evolution hones sense organs and brain areas to be exquisitely sensitive. Elephants, for example, can smell water from up to 12 miles away, while cheetahs can spot potential prey up to 3 miles away.

Echolocation requires more than just hearing; it also requires the bats that use it to produce sound waves that bounce off objects before they detect them. Their brains then produce a form of map in reflected sound that the bats use to spot insect prey. Although producing sound to detect food is relatively unusual in the animal kingdom, it is not a trick unique to bats. The aye-aye is a lemur, a species of primate, that is found only on Madagascar. It survives on insect larvae it extracts from tree trunks. It finds them by tapping a specially adapted finger on the trunk and listening with its oversized ears until it locates a cavity underneath the bark where a tasty morsel may be hidden.

Once the food has been detected, it needs to be acquired. Some herbivores have it easy, with the bison needing to simply put its head down and graze. In contrast, wolves need to chase down their prey before killing it, while orang-utans need to track down fruiting trees, pluck fruit from them, and sometimes peel the fruit, before consuming it. Killing prey can be dangerous, with buffalo and giraffe sometimes killing lions, and bison mortally wounding wolves. All animals have adaptations to help them acquire their food. Having detected a cavity within a tree, the aye-aye uses its forward-pointing incisors to drill a hole into the trunk before using its significantly elongated middle finger to prise the insect prey from out of its hiding place. Both the angle of teeth and the extension of a single digit on each hand are unique to the aye-aye, being found nowhere else in the animal kingdom. These phenotypic traits have evolved to enable the aye-aye to thrive. Aye-aye ancestors with these traits succeeded in avoiding death and failure to reproduce by detecting, acquiring and using food, and the phenotypic traits, and the genes that determine them, spread through the aye-aye population.

Even if acquiring food can be relatively straightforward, it needs to be done without becoming food for something else. The antelope on the African plains needs to be vigilant for cheetah, lion and wild dog, while the mouse searching for seeds, fruits and insects in the leaf litter must do so while avoiding snakes. The impressive 100 km/h top speed of the pronghorn antelope, the colour-changing skin of the chameleon and the hedgehog's resistance to snake venom are all adaptations to reduce the risk of death while foraging. Evolution has been inventive, with arms races between predators and prey resulting in remarkable camouflage, staggeringly fast top speeds and senses honed to detect the minutest smells.

Once food has been acquired, it needs to be utilized. Things can now become more challenging for herbivores like the bison, which need to ruminate. They initially swallow the vegetation they consume, before regurgitating it and further chewing it. As they chew their cud, they release an alkaline saliva that counters the acid in their first stomach, the rumen. They then swallow their twice-chewed meal, which can be more easily digested by the chemicals and bacteria in their guts. A bison can spend up to eight hours a day chewing its cud. Energy from grass is hard won. Koalas arguably have it even harder. They survive on a diet of eucalyptus leaves that contain many protective chemicals to keep insects at bay, and this makes them extremely hard to digest. The koala has a longer gut than other animals its size to aid with digestion. The longer it takes for the chewed leaves to move from one end of the gut to the other allows more time for leaves to be digested and goodness extracted. Still, despite this adaptation, the koala sleeps for up to twenty hours a day, during which it does nothing but digest.

Once the food is digested the energy needs to be allocated to all the different tasks an animal must do. Some energy will need to be dedicated to reproduction, some to fighting disease, some to detecting and acquiring the next meal, and some to just keeping the metabolism ticking over. Some species of animals use energy sparingly, eking it out over a long period, while others use it more quickly, often living a 'live fast and die young' lifestyle.

The guppies in Trinidad where I work occur with predators in streams and rivers on the floodplains at the base of the Northern Range mountains. Guppies in these streams will inevitably be predated. Individuals do not live long, and the predators keep their prey population sizes small, meaning there is plenty of high-quality food for the guppies in the form of nutritious stream invertebrates. The guppies are well adapted to live in this environment. They reach sexual maturity at a young age, when they breed they produce large litters of small young, they have a very high metabolic rate and are able to swim quickly, and they are very vigilant for predators. Their mouths are adapted to suck prey out of the water column, and their shortish guts allow them to digest their nutritious food effectively.

If you follow one of the streams up into the mountains, you soon encounter a waterfall. Predatory fish are never found above these waterfalls, but guppies sometimes are, yet they are different from their relatives below the waterfalls. Because there are no predators, their population sizes increase until there is insufficient food to go round. Instead of dying by being predated, these guppies starve to death. Their high densities mean high-quality food is scarce, and the guppies switch their diet to include algae and bacteria. The predator-free, low-food environment selects for different phenotypic traits than the high-food, high-predation streams. Guppies evolve new jaw shapes to scrape algae and bacteria from rocks, they evolve slower metabolic rates so they use energy more efficiently, and their gut becomes longer to allow them to extract energy from their less nutritious diet. Because they do not need to sprint away from predators, their maximum swimming speed is slower. They grow more slowly than their high-predation cousins, reach sexual maturity at a greater size and age, they produce fewer, larger offspring, and their maximum longevity increases. In the absence of predators, males become more brightly coloured, and they compete more intensively for females. Numerous phenotypic traits change, and the fish, although clearly still guppies, differ considerably between predator-rich and predator-free streams.

These changes have occurred because different genes are switched on and off at different times during development.

Natural and sexual selection has moulded the guppy to thrive in environments with very different causes of death and reproductive failure. It does this by favouring sets of phenotypic traits that help guppies avoid an early grave and a failure to breed in contrasting environments, and as these traits evolve, different alleles on different chromosomes become favoured and high and low guppy populations start to genetically diverge. In time, perhaps, a new branch of the tree of life will emerge. The guppies are a microcosm of how evolution has sculpted life over the past 4 billion years.

In this chapter I have explained how competition for resources is the driver of evolution, and how on our limb of the tree of life it has produced complex multicellular organisms that sexually reproduce. I have also given a high-level summary of how animal life has emerged and evolved until mammals inherited the Earth. Along the way, I have summarized how evolution is a powerful process that has created strain MK-D1, *E. coli*, aye-ayes, guppies, wolves and you and me. Each species is a success story, surviving in a world where death or reproductive failure is never far away. The ancestors of each organism alive today have survived periods of food shortage, landscapes rich in predators, cold climes, and warm ones. They have lived in water and on land and have survived volcanic eruptions and meteor impacts. They have found many ways to thrive, some as single-celled wonders, others as complex animals consisting of trillions of cells. In complex animals such as you and me, we have evolved keen senses to detect, acquire and efficiently utilize a wide range of food, and phenotypic traits to allow sexual reproduction.

For mammals to exist, multicellularity, sex and a whole host of other attributes I do not mention had to evolve.

These characteristics all evolved in response to the environments that ancestral species on our branch of the tree of life experienced. Modern mammals are a product of billions of years of evolution of competition for resources in both cold and hot environments, as well as wet and dry ones, in places where competition was sometimes intense and sometimes a little more relaxed. At times the journey was driven by volcanoes, earthquakes and asteroid and meteor collisions, and at others by predators, disease and limited food. Sixty-six million years ago, mammals became the dominant animal species on land. They spread and diversified, and the next step on our journey is how they became human. But before I discuss that, there is an important phenotypic trait I am yet to mention: consciousness. And that is the topic of the next chapter.

Consciousness

If you were to encounter an intelligent alien, they might ask you what it is like to be human. I suspect they are much more likely to ask such a question than undertake the type of orifice examination reported by many who claim to have been abducted by little green men. Would the first thing you do on encountering a sentient species from another planet be to poke your finger into any openings it may have? Probably not. So how might you answer questions on what it is like to be human? You might talk about your body inhabited by your mind, where your mind is the part of you that reasons and thinks. You might say that your mind is what makes you you. That would be a sensible response given our minds feel special and unique, and for some people too good to be temporary as they believe their minds live on after death. Observation forces us to accept that when we die our bodies serve no function, and decay, but it can be harder to accept that our minds die too. We might fight the inevitable ravages of age on our bodies by eating healthily, exercising daily and avoiding guilty pleasures that doctors tell us are bad for us, but it can sometimes seem inconceivable that our minds will wear out and die. Alzheimer's disease and dementia demonstrate that our minds can, and sometimes sadly do, fall apart like the rest of our disposable soma, but these illnesses do not stop many people believing their minds will live on post death. Our minds feel special and a little bit different from our bodies.

One reason that makes the mind seem special is we have

been unable to locate it in the way we have the heart or spleen. Unlike a kidney, liver or lung, there is no organ called the mind that we can dissect out of the body and study. There is no mind transplant procedure, and even though science fiction writers describe futures where minds can be downloaded on to a USB stick and uploaded into a computer, this is not going to happen any time soon. Given we can't locate the mind, how could it decay along with our brains, skin and muscles? However carefully we dissect the human corpse, no mind has been found, so perhaps it moves on after death. Personally, I don't subscribe to this view. My mind resides in my brain, the organ (or more accurately the set of organs) found in my skull, and upon my death it will no longer exist. My brain, and therefore my mind, is part of my disposable soma and none of it has life after death. I die and the lights go out. Of course, this makes me sad, but it does raise another question: why do we have minds, why are we aware of our thoughts and existence? Is it simply the universe's way of marvelling at its own magnificence, or just an accident of nature? Can we study the evolution of the mind in the same way we can study the evolution of any other phenotypic trait?

The hard problem of consciousness, a term coined by David Chalmers in 1995, is concerned with why we experience anything at all, and it has been central to consciousness research for the last few decades. The problem is wryly named, because all aspects of studying the brain and consciousness are hard; there are no easy bits. But the hard problem was seen as particularly difficult. Put another way, it asks, would it not be possible for a conscious-less zombie to behave just like us, doing everything we do but experiencing nothing? The hard problem is hard because it assumes that even if we get to a point where we understand everything about how the human brain works and are able to build an accurate computer model of it, the question of why we experience life would still remain.

What it feels like to be human is important, and perhaps the most important part of our existence. Given this, it is no surprise that researchers have been keen to understand why being human feels like it does. An obvious follow-up question is what does it feel like to be a bat, cat, or rat? We know the structure of the brain of each species holds the key to answering this question, and we also know that being conscious is the process that gives us the sense of self. Understanding how our brain works, and how our brain differs from those of other animals, has helped scientists begin to understand consciousness and why we have minds.

An influential scientific paper on consciousness by the philosopher Thomas Nagel has the great title, 'What Is It Like to Be a Bat?' The paper starts by assuming that most animals have some degree of consciousness, takes the position that we can't understand consciousness by breaking it down into physical processes such as interactions between cells in the brain or the workings of genes, and concludes that it is impossible to know what it is like to be a bat, but, because bats are conscious, it must be like something. The paper was published in 1974, and much has been discovered since. Psychologists are now confident that bats do have a degree of consciousness, and these researchers also have a reasonable understanding of how a bat's brain works. But we still do not know what it feels like to be a bat. There are probably ups and downs, good times and bad, with bats experiencing a sense of pleasure and of pain. I doubt they feel anxiety about the fate of the insects they eat while pondering the pros and cons of vegetarianism (I'm assuming they're not fruit-eating or vampire bats), they are probably unaware that death awaits them, but they take actions to avoid predators, and although they communicate with one another up to a point, they do not lecture one another on the wondrous achievements of bats past.

Understanding consciousness is one of science's hardest challenges. Conducting experiments on consciousness is difficult, and most experiments conducted to date do not provide much insight into the hard problem of consciousness. Nonetheless, there are a range of views on why consciousness exists, often discussed by philosophers rather than consciousness researchers conducting experiments. When reading this literature, I had to use my mind to try to understand why it exists, trying to balance evidence against speculation. I have concluded that there is evidence that most animals have a degree of consciousness. A fly or a shrimp does not experience the world in the same degree of complexity as you and I, but they still do have experiences, and these experiences are a form of consciousness. Their form of consciousness may be very different from ours. Octopuses, for example, 'see' the world through their skin, and many octopus species can change colour and pattern in response to a change in light or background colour and texture. Working out what consciousness is like in other species is extremely challenging and perhaps impossible, and genetics does not help us much here. There is no evidence that there is a single gene that turns consciousness on or off either in octopuses, shrimps, flies or humans. Despite the horror film industry's love of mindless, murderous zombies, a mutant who lives their life just like you and me but without experiencing any of it has never been observed. There is no simple, single cause of consciousness, although we can use drugs to turn it on and off.

I am not persuaded the hard problem is an issue, but this will be seen as contentious by some researchers. Consciousness inevitably arises from the workings of brains, and I suspect it would be impossible for a brain to receive and process signals from the eyes, ears, nose and skin, combine this information with memories and decide how to act without experiencing something. Given we are conscious, and we have been unable to find what makes us conscious, the logical conclusion is it must arise from the way that brains work. We do some things subconsciously, of course, but this does not mean that everything we do can be done without us knowing about it. The starting assumption of the hard problem that everything could be done subconsciously is wrong. Zombies that behave just like conscious people don't exist because they can't. A growing number of psychologists are coming to the conclusion that conscious minds are an inevitable consequence of a complex brain, and if we can understand how the brain works, that could reveal why we are conscious.

Science is making great inroads into understanding our minds, and much insight has emerged from understanding the workings of our brains. Perhaps in time evidence will emerge that will show that being able to move in response to an event that an organism senses in the outside world requires the ability to experience, and this underpins consciousness, but science is not quite there yet. I suspect that not only will we fail to find zombies, we will also fail to find zombie chickens, zombie shrimp and zombie dogs. There are lots of publications on zombie insects, but these are animals whose brains have been infected by the fungus *Ophiocordyceps unilateralis*, the pathogen that created human zombies in the fictional TV series *The Last of Us*. In real life, the fungus kills infected ants, flies and spiders, and before they die it alters their brains in such a way they seek out a final resting place that maximizes the chances of the fungus infecting other insects. Whether such an infection alters the consciousness of insects, if indeed they have any, is unknown. Consciousness is an evitable feature of the brains of some animals, including us, and it emerges because of the way the brains work. How conscious an organism is is thought to vary with the degree of complexity of parts of the brain. There is still a lot to learn about consciousness across the corner of the tree of life where animals are found.

Nagel's question about what it is like to be a bat made me think about what it is like to be me. It is a surprisingly difficult question to answer, in part because I don't have any comparison about what it is like to be anything else other than my younger self. Nonetheless, on balance I think being me is OK. The senses I use to experience the world – sight, hearing, touch, taste and smell – all work. My eyesight has never been particularly good and, as I've got older, I've had to buy reading glasses, but my poor vision hasn't stopped me from experiencing many remarkable things. Similarly, my hearing isn't as good as it used to be, but overall, I am still able to experience the world, and I get enjoyment from seeing nature, from listening to music, from human contact, and from good food and drink. The organs I have to experience the world are functioning.

My memory, something that is a key part of consciousness, is also functioning adequately. I have a good memory for facts, particularly about biology, and when I reminisce with friends and family I can recall past events as well as they can, even if sometimes I wish I could forget daft things I have done. My memory for faces is less good, but it has always been thus. When acquaintances used to greet me on the street, and we would have a conversation for a few minutes, Sonya used to ask me why I didn't introduce her. The answer was always the same: I had no idea with whom I was chatting. She no longer asks the question, simply observing that I had no clue who they were. It is the same with famous faces. Much to my children's chagrin, I had no idea I had been at a drinks event also attended by the singer Katy Perry until someone told me who she was. On the plus side, she didn't know she was at a drinks party with me either.

Deciding how to act in a situation is also part of consciousness, and in pondering what it is like to be me, I thought about how I make decisions. I am good at making decisions, even if they end up being bad ones. I try to base all my decisions on information available to me and, having made them, I don't worry about the decisions being less than perfect. If further down the line I wonder if I could have made a better choice, I try to learn from my mistake, thinking through where I made an error, but I do not get anxious about it. I tend not to panic, and I rarely get stressed by situations. Shortly after my brush with malaria in my youth, I got quite anxious about my mortality, but eventually realized there was little point worrying about it. I developed a strategy where I divide situations into those that I can control (my choices), those that I may be able to influence (other people's decisions), and those that I have no impact on. I don't stress about the latter two as what's the point, and I am in control of the first. I do think carefully how I might influence other people's decisions by working out what they will want from a situation, where our interests align, what leverage I might have, and what might appeal to them. This is part of being human, and something that all of us do.

As well as being me, I have also spent time trying to work out what it is like being Woofler. It is clearly quite different to being me. He is well equipped to experience the world, with excellent vision, hearing and smell. He also has an active memory, particularly of places where he has had success hunting squirrels and rabbits, but he also remembers other dogs and people, and some of these memories are long-lasting. His decision-making often fails to meet with my approval, but I do believe that not all his actions are pure instinct. He will not take food from a plate when someone is in the room, but as soon as they leave, he will. Woofler does not have the wherewithal to understand that by a process of elimination we can work out that it is he who stole the Man-chego. He is also not allowed on our bed, and will not jump on it when we are in the house, but if we go out he makes a nest among the pillows and we hear the thump as he jumps down when we return home. The dog will also often look at me, perhaps trying to assess my likely response, before decid-ing on whether to do something that he knows will lead to me telling him off. This is usually having a go at other dogs that are normally sufficiently large they can beat him up, or rolling in fox poo. On some occasions he will walk on by, but often he appears to decide that getting in trouble is worth it. I am, of course, anthropomorphizing Woofler's behaviour, but if we could measure consciousness, I am confident we would conclude he has a degree of it.

It turns out we can measure consciousness. Doctors do this by using brain scanners that detect patterns of electrical activity within the brain. Brain scans can detect electrical fields produced by large groups of brain cells in different parts of brains. Scientists have come up with ways of describing these electrical patterns by placing them on a scale from purely random through to highly structured and entirely predictable. The electrical activity of a waking, conscious brain when you are thinking about the challenges of daily life lies between these two extremes. If you could scan my brain as I write this paragraph, you would see electrical waves beginning in one part of the brain that go on to move to other areas, before fading out. The electrical activity in my brain is not random, with cells flashing on and off in an unpredictable way. The electrical patterns are also not highly predictable. It is not possible to say with confidence which cells will next become active because nearby neighbours are active now. Signals instead propagate in a particular location before moving as a wave, but not in a way we can always accurately predict. The waking brainwaves of a normal person that the hospital scanner detects are organized, but not rigidly so.

Our brain's electrical activity changes when we do different things. When you initially go to sleep your level of consciousness changes and your brainwaves become more predictable and more localized. They slow down even further when you fall into a deeper sleep, before speeding up again later when a type of sleep called rapid eye movement, or REM, begins. Mind-altering drugs such as LSD also change the state of consciousness, with the electrical pattern in the brain becoming more random. The brain scans of anaesthetized people show even greater randomness than those on mind-altering drugs, with consciousness disappearing completely. Being anaesthetized is very different from being asleep, in that after you have slept you are aware that time has passed. If you are anaesthetized, being put under and waking up are experienced as instantaneous, and during that period your brain's electrical activity is disorganized and random for much of the time. It is like being in a trauma-induced coma.

One of my children experienced this when she was twenty. She caught bacterial pneumonia and subsequently developed sepsis. She was put into a drug-induced coma for two and half weeks while the fabulous NHS staff worked to save her life. We were briefed that when she was brought round, she would have no idea how long she had been unconscious. The doctors were right: she could not believe she had been in a coma for so long and found the whole experience disconcerting. I am delighted to say she has made a complete recovery. She was almost the same age as I was when I had malaria, but I was lucky I lost only a few minutes of my life, compared to the seventeen days my daughter has no memory of. If we had monitored her brainwaves during her coma, they would have been random. Our levels of consciousness vary depending upon all sorts of things, and even what we are doing, with each state yielding a different pattern. Conscious thought has a particular pattern, while subconscious processes have another. Characterizing these different brainwave patterns is an active area of research. One reason psychologists think that some species of some animals have consciousness is they also have brainwaves that are neither entirely random nor highly uniform. For example, in 2023 an international team of scientists described how tiny implants were used to monitor the brainwaves of three free-living octopuses. The brainwaves detected when learning new tasks were very similar to patterns seen in mammals, including humans, while other behaviours generated patterns that were like nothing seen before in any other animal. Research such as this across a wide range of animals reveals that some types of brainwave patterns are quite similar, even between distantly related species.

A few neuroscientists have gone so far as to state that consciousness is the brain's electrical activity, although most don't subscribe to this view. Electrical activity certainly provides the best approach we have for measuring how conscious someone, or something, is, but it does not provide an explanation for consciousness. The electrical pattern arises because of the way that brains work, and biologists, neuroscientists and psychologists are making great progress in revealing their inner workings. Some neuroscientists have argued that brainwave patterns in other species, including houseflies and shrimps, suggest that a degree of consciousness is widespread across animals, but we do not yet know what it feels like to be a housefly, Woofler or a bat.

The part of your brain where consciousness resides sits inside a dark box, your skull. It does not detect light, smells, sounds, touch or taste, but it turns electrical signals from cells in organs that do detect these things into experiences. The brain inside our skull receives signals from the eyes, ears, nose, mouth, skin, muscles and other organs as either chemical signals or electrical currents, and the brain then makes sense of these before sending chemical or electrical signals back to our muscles and other organs instructing them what to do. Some of these instructions are sent subconsciously, others following a decision having been made. All this can happen astonishingly quickly, if need be, but some decisions can take an age. Sonya can take minutes before deciding between the hazelnut noisette and the orange crunch in a box of Quality Street chocolates. She usually makes a mistake regardless of her choice, handing me a half-eaten chocolate, and much more quickly makes a second decision to eat the other variety to correct her initial error.

Your sense organs contain cells that can detect aspects of the external environment. Your retina, an organ at the back of your eye, contains nearly 97 million cells that can detect light of different colours and intensities. Each time particular proteins in the cells are hit by a photon they create an electrical charge, information that the brain can use. Similarly, cells in your nose contain proteins that attach to different molecules, allowing you to detect smells. When you cook bread the action of yeast in the dough produces chemicals called ethyl esters. Some cells in your nose have proteins that can bind to ethyl esters, and when they do, they send an electrical signal via the olfactory nerve to your brain. You then decide what to do, which in my case is to usually search the fridge for some butter. Your other sense organs operate in similar ways, but instead of detecting light or smelly molecules they instead detect sound, touch or texture before sending electrical signals to the brain.

The brain takes all these signals from the sense organs and builds a simulation of the world around you inside your skull. You might smell a freshly baked bread roll, locate it by looking around the kitchen, and determine it is still warm by touch. Your brain simulates an experience of finding a freshly baked bread roll and you then decide what you are going to do with this information.

The simulation of the world our brain creates may, or may not, be accurate, but it is sufficient for us to survive and hopefully thrive. There will be differences between the simulations that different people's brains create. For example, my sense of greenness, or yellowness, or redness, may well be different from yours (this is the hard problem of consciousness again), but it matters little if we can both agree that an object is a particular colour. The important thing is that the simulation of the outside world we experience contains sufficient information for us to decide to take sensible actions. For most animals, these actions involve the detection, acquisition and utilization of resources such as food, water, a territory or a mate without succumbing to something that could kill them. In humans our actions go beyond this, as they might involve completing a crossword, taking the dog for a walk or deciding what colour to paint the bedroom.

Building a simulation of the world, and working out how to respond to it, is a complicated task. The human brain is the most complex object in the known universe. It is amazing, so look after it. Your brain is your body's command centre, and its primary role is to keep you alive. Some of what our brains do happens without our knowledge. It moderates our body temperatures, manages our heart rate and controls many instinctive behaviours. These all happen in the subconscious, although with concentration we can take conscious control of some of these processes, such as our heart rate. As well as doing all this, our brains also store memories, and these play an important role when we decide how to act in a given situation.

The human brain is divided up into many separate parts, but at the crudest level it can be partitioned into three distinct areas: the brainstem, the cerebellum and the cerebrum. The brainstem is evolutionarily the oldest part of the brain and it regulates most of the subconscious processes that are required to stay alive. It keeps the heart pumping, the lungs breathing and many of our other organs functioning. It is found at the base of our skull by the top of the spinal cord. The second area is found towards the back of our skull and is called the cerebellum. It is the command centre for movement, and is how we voluntarily control our muscles. As I type this paragraph, it is the cerebellum that is directing my fingers where to go on the keyboard and when to press down as I type. The cerebellum is all about fine motor control, and if it were to be damaged my typing would become wonky. But that is not all: this part of the brain also plays a role, along with another part of the brain, the amygdala, found in the cerebrum, in regulating our responses to fear and pleasure. The third area of the brain, the cerebrum, is the largest part and it is responsible for sensing smell, speech, thought, emotions and learning. The largest part of the cerebrum in humans is the neocortex, and this organ alone takes up over half the size of our brains. It is the neocortex that gives us our intelligence and likely our heightened sense of consciousness.

Each of these areas of the brain can be further divided, usually by the function it performs, with each having a name. You need a good memory to be able to keep track of each named area of the brain, what each does, and how it connects to sense organs and other parts of the brain. The intricate ways in which signals are processed and combined to create a simulation of the world are complex and fascinating, but the details are beyond this book. What is central to the brain is the network of connections between brain cells called neurons, and although these networks operate in slightly different ways in different areas of the brain, there are broad similarities in how the networks function.

Your brain consists of between 80 and 100 billion neurons, and at least as many, and perhaps many times more, glial cells that look after the neurons. In the nineteenth century, scientists thought that without glial cells the nervous system would fall apart, so they named them after *glia*, the Greek word for glue, but that is not what the cells do. Instead, glial cells are to neurons what a team of mechanics are to racing cars, ensuring they are in good working order. The ratio of glial cells to neurons differs between different parts of the brain, and there are highly variable estimates of how many of each type of cell there are. Counting the individual cells in a human brain would be a laborious and not hugely exciting endeavour, so we rely on approximate estimates obtained by scaling up counts from small sections of the brain. Regardless of their number, glial cells are fundamental to brain function, but, as far as we know, their role is not critical to understanding consciousness, and for that reason I do not consider them further.

There are lots of different types of neurons, but all contain the same key components. They have evolved to be able to form large networks of interconnected cells. A single neuron consists of a main cell body, called the soma, where the cell nucleus and other organelles are found. Biology can sometimes be frustrating with its terminology. Soma means body and comes from the Greek *sōma*. The 'disposable soma' refers to our bodies, while the soma of the neuron is the main body part of the cell. Extending from one end of a neuron's soma there is usually a long, thin tube called the axon. It is covered in a particular type of protein called a myelin sheath that allows it to rapidly transmit an electrical signal. At one end of the cell, typically closer to the soma, is a complex branching network of tubes called dendrites. These look a little like the branches of a tree, in that there are many more tips than there are dendrites leaving the soma. The dendrites are covered with synapses, such that there can be many thousands of synapses on a single neuron. A synapse is the way neurons communicate with one another, a link between cells. They usually link to other neurons, but they can also link to other cells such as those that detect light in your retina. The axon also branches, and at the end of each branch is something called an axon terminal. These terminals form links to dendrites of other neurons via synapses. Each neuron can link to thousands of others, with dendrites of one cell forming synapses with axon terminals of another. The large number of synapses on each neuron allows neurons to create a vast network of interlinked cells, with there being about 600 trillion synapses in your brain. The network is fundamental to how you function, and to consciousness as well.

We have all become familiar with the concept of networks

Structure of Neuron Cell

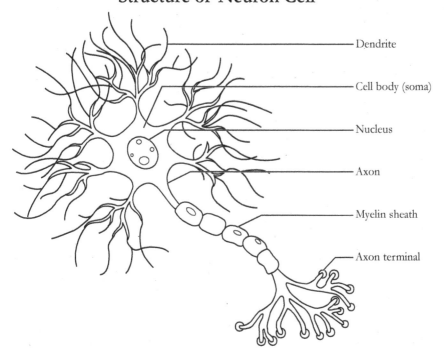

- Dendrite
- Cell body (soma)
- Nucleus
- Axon
- Myelin sheath
- Axon terminal

in recent years thanks to online platforms like Facebook and X. Each user of a social network platform forms a set of connections to their friends, acquaintances or people that interest them, and each of the people they connect with has their own set of connections to other folk. In scientific parlance each person is a node, and each connection is an edge. In the network of neurons in the brain, each neuron is a node, and each synapse is an edge. Social networks extend across the planet, and it is claimed that if you were to draw a network of the entire human population, you would find that any two random people are linked via six degrees of separation. In other words, I could be linked to you via five people who are acquainted with one another. I will know one of these people, you will know another, and the remaining three will complete the chain. There will be a few exceptions to this rule such as the isolated tribes on some of the Andaman Islands, but the six degrees of separation will likely be true for the readers of this book. The network of neurons in the brain is even more connected. There are estimated to be between three and four degrees of separation between neurons in the brain, which should give you a hint that the network in the brain is highly connected. It is also extremely large, given there are about fifteen times as many neurons in your brain than there are people on the planet.

The social network analogy is useful in providing a little insight into how the brain works. Imagine there are six Instagram accounts devoted to describing what I am wearing and what I am up to. One account is entirely devoted to the shoes I wear each day, a second to my trousers, a third to my shirt, a fourth to where I am, a fifth to what I am doing, and a sixth to what I am saying. These accounts are so unbelievably dull that no one could follow all six without passing out from boredom, but let's assume that each of my three children follows two accounts. Luke is a devotee of my footwear and trousers, Georgia is interested in my shirt and where I am, and Sophie keeps track of what I am doing and saying. Because I am not very good at keeping Sonya informed, she follows each of the children. On the rare occasions I have a garish new shirt, some snazzy new cowboy boots, am doing something interesting, or have said something of note, Sophie, Luke and Georgia share my posts and Sonya sees them. Sonya consequently acquires useful information from across the simple network and can then decide how to act accordingly, perhaps by suggesting the shirt does not match the boots.

Information flows through the vast network of neurons in the brain in a related way. To understand how this happens, information on how neurons work is helpful. Each neuron operates by receiving inputs from synapses that are then sent as electrical signals that travel along the dendrites to the cell body. Depending on the number of synapses that fire and the number of signals received from the dendrites, the soma may then send an electrical signal along the axon. When this electronic pulse, or action potential as it is referred to by neuroscientists, reaches the axon terminals, in most neurons it results in the cell releasing chemicals known as neurotransmitters. These chemicals bind to receptors on the dendrites on the cell that forms the other side of the synapse. If this is another neuron, the neurotransmitter may result in an electrical signal being sent along the dendrite on its way to the soma. If the cell on the other side of the synapse is a muscle cell, it may contract, or relax. In some cases, electrical ions are passed between axons and dendrites rather than neurotransmitters, but the result is the same – one cell sends a signal to the other. Synapses using electrical ions are faster than those that rely solely on neurotransmitters because electrical currents move faster than molecules of neurotransmitters, but the type of signal that can be transmitted is more restrictive, essentially being just a yes or a no, while a neurotransmitter can convey many degrees of maybe.

The light-sensing cells in your eyes each connect to multiple neurons called retinal ganglion cells. Each one of these neurons connects to about a hundred light-sensing cells, but each does so in a different configuration. Some neurons connect to light-sensing cells scattered across the retina, while others are connected to light-sensing cells in only a small part of it, with yet more connecting to columns or rows of light-sensing cells. The retinal ganglion cells form synapses with large numbers of other neurons, with these neurons receiving information about patterns of light intensity and colour, along with edges and shapes of objects. In turn these neurons form synapses with many others, and with each link a more specific picture of what you are looking at is generated. Eventually some neurons deeper in the network will be triggered that are responsible for letting you know you're looking at a person, or an animal, or a tree, or a lake, or whatever. In one study, researchers identified a neuron in a subject that fired every time he saw a picture of Jennifer Aniston. If you know what the actress looks like, you too will have neurons like this. In fact, you may have several Jennifer Aniston neurons. You will have several neurons that can fire if they are looking at any familiar object.

The networks of neurons in your brain consequently combine information, with each layer building up a more detailed picture of what you are experiencing. Information can also be combined across the senses, with links between

A Network of Neurons

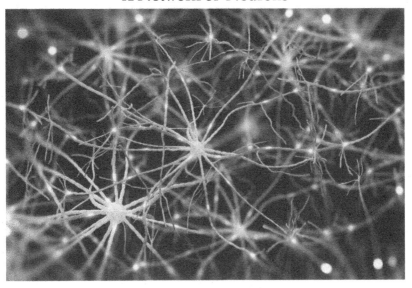

neural networks devoted to working out what you are touching, smelling, looking at and hearing. As your networks of neurons build up a picture of what you are experiencing, the brain also filters out redundant information. A neuron only fires, sending an electrical current down the axon and a signal on to other cells, if enough synapses are triggered. What this means is if there is very little information to suggest you are looking at a picture of Jennifer Aniston, then your Jennifer Aniston neurons will not fire.

There is significant overlap in what neurons do within the brain. Several neurons in a part of the brain called the hippocampus could fire if you see a picture of Jennifer Aniston. Other neurons might also fire that are in some way associated with Jennifer Aniston. For example, a neuron linked to Courtney Cox, who, along with Aniston, starred in *Friends*. This suggests that some neurons may code for the abstract concept of *Friends* rather than just one of the actresses. What to do when different neurons fire that encode different information? This response has been studied in detail in another part of the brain, the neocortex.

Much of the heavy lifting your brain does in terms of identifying patterns happens in your neocortex. It is folded in complex ways, and is typically between 2 and 3 millimetres thick. If removed and laid out flat post-death, your neocortex would cover a square with sides of about 45 centimetres. Via careful measurement with high-tech, high-powered microscopes, researchers have estimated that there are 100,000 neurons within each rice-grained size volume of your neocortex, linked by half a billion synapses. These neurons are arranged into between one and two million pillars known as cortical columns. In humans, each of these columns contains a little over one hundred neurons, arranged into six layers.

The majority of neurons in the neocortex are arranged in a plane running from the skull towards the centre of the brain. Other neurons are oriented across this plane. These neurons shuttle information between different parts of the neocortex. Perpendicular neurons somehow appear able to achieve a consensus on what you are experiencing. For example, if you are looking at a picture of Jennifer Aniston, and twenty Aniston neurons fire and only three Courtney Cox neurons fire, the consensus reached will be that you are looking at a picture of Jennifer Aniston. The details on exactly how this is achieved are not yet fully understood although progress in neuroscience is currently so rapid it likely won't be long before we understand it better.

Another area of active research is how we lay down memories. You must have seen pictures of what you are looking at to know what it is. We don't instinctively know what something looks, smells or tastes like, but instead we need to learn. We do this as we experience things. For example, most of us have never met Jennifer Aniston and have instead learned who she is from watching her act in blockbuster movies or TV shows. We learn the faces of our friends, acquaintances and work colleagues through meeting and talking with them, and synapses form that allow us to recognize them. Although scientists have yet to work out the details of how new memories are formed and stored, and old ones are lost, it is clear that sleep is when much of the work is done. Getting a good night's sleep really does help you remember.

I have so far described how the brain identifies pictures or smells, rather than a dynamic series of events, such as a movie or a smell getting worse. I have explained how the brain interprets a picture of Jennifer Aniston, but not what she is doing in the film *Marley and Me*. The brain is very good at identifying stills, but it can do so much more. To survive in the real world, we need more than a series of static snapshots. Instead, the simulation our brain produces needs to be dynamic. It is more like a video game than a picture postcard, and it needs to predict. To create a dynamic simulation of the world, the brain needs to monitor moving objects, either visually, auditorily, via touch or smell, and to predict where they will be in the future. The way the brain does this is a little bit like the autofocus tracking feature on modern-day cameras that can maintain focus on objects as they move across the camera's viewfinder, but rather than tracking just one object, the brain can track many simultaneously, can work out the distance between them, and can predict the consequences of these objects being rotated in three dimensions.

The way that the neocortex creates the dynamic simulation of the world in our head is described by something called frames of reference theory. A frame of reference is a way of using coordinates to describe the location and movement of objects, and is most intuitively understood by focusing on vision, although the logic can be extended to hearing, touch and even to language and complex thought. Frames of reference are used to describe the origin, orientation and distances between objects, and their rates of change with time. The origin is the central point, and in the case of the brain, it is you, the observer. The orientation describes how objects are rotated with respect to one another. For example, if you are upright and are standing in front of a clock face looking at the hour hand, it appears vertical and pointing upwards when it is noon, vertical and pointing downwards when it is pointing to the number six, and horizontal and pointing left or right when pointing to the nine and three numerals respectively. If you rotate the clock face, in any direction you wish, the orientation will change. Our brain is excellent at tracking the changing orientations of objects.

The distance between two objects allows the brain to work out how close or distant they are from one another. The distance is usually thought of as being between the observer and the object, but a frame of reference system can be used to work out the distance and orientation between any two objects, regardless of whether one of them is at the origin or not. You can tell when two cars are on a collision course with one another, and can predict when the crash will happen, by repeatedly tracking the distance to each car, the distance between each car, and the fact they are on a course to collide.

A frame of reference describes how objects move relative to one another and relative to you, the observer. The brain monitors how the orientation and distance between objects change with time. Our neocortex is remarkable as it can simultaneously work out frames of reference for multiple objects in our fields of sensory perception. It does this via large numbers of synapses firing as they are stimulated by a particular pattern, but also via using what are known as grid and place cells to construct a map of the world around us. Our remarkable brains can take all this information and construct a simulation of the world and predict what is about to happen.

Once a simulation of the world has been created, the brain uses it to predict what will happen next – how each object with a reference frame is expected to move. If new information from the sense organs matches with predictions from the simulation, then new signals from the sense organs are not cascaded far into the network but instead die out. We don't experience them. In contrast, if the predictions of what will happen and observations diverge, then the simulation needs to be updated, often by drawing on information from neurons and cortical columns that had perhaps not recently contributed to building the simulation.

The way the brain compares predictions with observation is clever. For a neuron to fire, sending a signal down its axon and on to elsewhere in the network, it requires several synapses to fire, and for neighbouring neurons in the cortical column to be in the correct state. For a neuron to fire an electric signal down its axon, its soma needs to receive sufficient electrical signals from multiple synapses to push it over a threshold. If that threshold is not reached, the neuron enters a short period of time in a state where it is ready to fire if it receives another nudge. Neighbouring cells can detect this excited state, and this can prevent them from firing, a phenomenon known as lateral inhibition by neuro-scientists. A consequence of this is that multiple neurons in a cortical column only simultaneously fire, sending signals throughout the brain, if they all experience many synapses firing simultaneously. This might happen if the simulation and new observation diverge – i.e. the simulation of the world is inaccurate.

Once a simulation of the world is running within our brains, the body may need to be directed to respond to events to remain out of harm's way, or to acquire a resource. Some responses are entirely instinctive, such that an event in the simulation always results in a particular action. An example of such a response might be when a particular odour is sensed within the environment, and you always move away from it. Alternatively, we can use past experiences to decide on how to respond to a particular event in the simulation.

The area of the brain that brings together past experi-ences with the brain's simulation of the world is the hippocampus. Like many scientific terms, the name comes from Greek. If you squint a little while looking at a diagram of the brain, the hippocampus takes the shape of a seahorse, which is the animal after which it is named. Neurons for memory can also run the simulations in our brain, and these are linked to networks that help you make a decision on how to behave. As with much of neuroscience, the details of how this happens are yet to be fully understood, but networks of memory and simulation neurons help us to make a decision. Once a decision is made, we act upon it, with signals sent to muscles about how we should move, or what we should say.

Consciousness researchers break down the steps from becoming aware of something in the outside world with our senses to responding to it as perception, attention, evaluation, integration, decision-making and action. From perceiving a ball thrown towards you, to moving your hands to catch it, happens quickly. Although the human brain is extraordinarily complicated, it is also remarkably efficient. But do our brains make good decisions? And how can we tell?

There is a whole field of scientific research about how to make decisions. Scientists working in this field research not only humans but also animals, and even artificial intelligence. You might be quite surprised at some of the insights that have been achieved. For example, consider the marriage problem, which asks what the best strategy is in choosing someone to marry. The problem has been attacked with computer simulations, but, as is often the case, some simplifying assumptions are needed to formulate the problem in such a way that a computer can help solve it. The marriage problem assumes that someone looking for a life partner dates prospective spouses randomly but sequentially, never two-timing a date. Everyone who is dated can be ranked, and there are no ties: for example, second place cannot be shared by two prospective partners. However, the person looking for a spouse does not know the ranking in advance. The problem also assumes that once a decision is made on whether to marry or move on, the decision is irreversible, such that the jilted person will be so hurt they will never agree to a second chance. Finally, assume that any marriage proposal will be accepted. Given these rules, when should you stop looking and settle for someone you like? Should you marry the first date you like, or should you keep looking just in case there is someone who would be a better life match? What is the best strategy to end up with the number-one-ranked partner?

The answer, it turns out, is a little involved. It depends

upon the number of prospective dates, which of course in reality is something we probably do not know. The solution to the problem is also couched in terms of probability. It is not something like you should always choose the third date that you meet. Rather, it is based on the probability of choosing the number-one-ranked applicant if there are a particular number of dates and you have already rejected a certain number. For example, if there are six potential dates and you marry the third you meet, then there is a 42.8 per cent chance you chose the best option. There are stories, possibly apocryphal, of decision theorists using algorithms like this to select their life partners, but it is not how I decided to ask Sonya to marry me, and she assures me it is not how she reached the decision to say yes. We married because we make each other happy.

When decision-making theory is applied to animals, it is used to address questions such as, when should an individual move from one location to another? The animal is assumed to maximize something tangible, such as the amount of food it can eat in a certain location. When the returns of remaining in a patch are lower than the benefits of expending energy to move to another patch, the animal should move. The problem can be made more complicated by including social groups of animals, varying degrees of knowledge about likely food availability in other patches, and how different areas might vary in the risk of being predated. The application of decision-making theory to animal behaviour has provided interesting insights into how animals should behave, and sometimes they do seem to broadly follow the optimal strategy. However, working out what motivates an animal is not always straightforward. Not all animals spend all their time struggling to find food. They can be motivated by finding mates, water, shelter, warmth, safety, or even knowledge about their environment such as where competitors might be, and these motivations change with age, season and the weather. Despite the complexity, many decisions animals make appear to be optimal. They make many very good decisions.

Humans are animals, so it is not surprising that decision-making theory has also been widely applied to understand, and predict, how we behave. One difference between us and most animals is our ability to think about our long-term futures, and when we plan for these we try to make decisions that we think will make us most happy. Are our brains trying to maximize happiness? Happiness can be a bit of an elusive goal. Within a population, there are differences in how happy people are, and about a third of these differences can be attributed to genetics. Some people really are born happier than others. However, perhaps as much as two thirds of these differences are due to our circumstances and how we live our lives. Being financially secure helps with happiness. People living in poverty tend to be less happy than those who do not, but being super-rich doesn't guarantee happiness either. Surveys of happiness across people on rich lists that ask participants questions about their state of well-being reveal they are only slightly more likely to be content than those who are not on the lists.

The problem with money is the more of it people earn, the more they tend to spend. A pay rise consequently can often result in only short-term gains in happiness, because the extra cash is often spent on nicer clothes, better food or more exotic holidays, resulting in concerns that the more expensive lifestyle cannot be maintained. A lottery win, salary increase or unexpected inheritance from a distant relative does increase happiness, but not for as long as you might expect. Happiness is much more than money. Individuals seem to have a baseline level of happiness that they return to. Increases or decreases in happiness (excluding depression) tend to be short-lived. Psychologists refer to our tendency to return to our baseline level of happiness with a great phrase: the hedonic treadmill.

People have a tendency to compare themselves with their friends and colleagues. These comparisons can become obsessive, and when they do, they usually result in a decrease in happiness. One reason why social media can be pernicious for some people is that the steady stream of apparent success in Facebook, Instagram or X feeds reduces happiness through leading to feelings of inadequacy. I deleted my Facebook and Twitter accounts several years ago, and although I had withdrawal symptoms for a few days, I have not missed them. I will likely need to sign up again to social media to promote this book, but the thought doesn't fill me with joy – even if any extra income from doing so buys me a couple of weeks of happiness.

Comparing oneself to one's peers and friends can reduce happiness, but having a strong friendship circle is healthy. Humans are a social species, and strong social bonds are important for happiness. This starts at home, with people whose partner is also their best friend tending to be happiest. I can relate to this. Sonya is my best friend, and that does make me happy. When she travels, I miss her and I am not as happy as when we are both in Oxford. But contact with other friends is important too. The absolute number of friends does not seem to be hugely important, rather the strengths of some of the friendships are key. Having one or two close friends can lead to greater happiness than having many tens of friendly acquaintances.

Friends are important for happiness, but comparison to them is not, and these observations underpin a strategy that appears to help improve happiness – keeping a gratitude diary. Writing down things you are grateful for can help quash negative emotions brought about by worrying you are not doing as well as your friends or colleagues are. Positive thoughts do improve happiness. If you keep a gratitude diary, you shouldn't write entries in it too often. Turning to your diary more than two or three times a week can lead to stress about not having enough to be grateful for.

A final thing that does seem to contribute to happiness is a sense of fulfilment in work. If your job is not a chore, then the happier you will be. I can relate to this following a series of unexpected events that happened to me at work since moving to Oxford in 2013 that, surprisingly, made me happy.

Part of my Oxford employment contract stated that, if asked, I would have to take on the role of head of department. I was assured when I was negotiating terms that this would not happen for very many years. However, by 2016 I was being suggested as a possibility for the next head of the Zoology department. So too was Ben Sheldon, a good friend and colleague. Neither of us wanted the job, but one of us had to do it, and it was up to the faculty to decide. Ben and I fought a campaign to lose the vote. Ben's campaign strategy was to do nothing, while I actively campaigned for him, and I was delighted when he was appointed department head and I his deputy. I knew I was next in the firing line, but I would have five years in which to write this book. Or so I thought.

Five months later, the building which housed both my department, the Zoology Department, and the Department of Experimental Psychology was shut with twenty-four hours' notice due to airborne asbestos levels crossing a health-and-safety executive threshold. The risk to users of the building was minimal, but the closure was a crisis for the departments involved, and the university. The Tinbergen Building had been the University of Oxford's largest research and teaching edifice, with 1,650 people regularly using it, and its closure created a major headache. It very quickly became clear to both Ben and me that dealing with the repercussions of the closure was going to occupy our lives for months, if not years.

The university organized frequent crisis meetings to work out what to do. These were chaired by the head of the university's administration and were attended by about twenty people, mostly senior administrators. It very quickly became clear to me that the university wanted to find a way to solve the problem of two homeless departments as painlessly as possible, and ideally without spending a fortune. I realized they would likely sort out a short-term solution of where to house those displaced by the building closure quite quickly, but that a longer-term solution was going to be much harder as there were no empty suitable buildings. This posed a huge threat and a massive opportunity for the department, and Ben asked me if I would lead on a strategy to dodge the threats and realize the opportunity.

I liked this challenge, and although I had never much liked the idea of putting my research on pause, I found that taking on the strategic role was rewarding. I felt I was achieving something. I developed, with input from many others, a vision and financial case for Biology at Oxford. It involved delivering an exciting new undergraduate course we had already begun planning, merging the Department of Zoology with the Department of Plant Sciences to form a new Department of Biology, growing the new department by about 25 per cent, and persuading the university to construct a building to house the new department, along with the Department of Experimental Psychology. Although Ben and I had no training on how to do this, and largely made it up as we went along, we had success in selling the vision to a few key people who encouraged us to develop it further. I twigged that to deliver the vision, we needed to sell it across the university's decision-making committees, and that involved me standing in an election to become a trustee of the university. I was elected following a ballot, and I became a member of the University Council, equivalent to the board of trustees.

The vision was accepted by the university, which agreed to a new building. We launched the new course, the merger date was set for 1ˢᵗ August 2022, and the department grew. However, becoming a member of the council had a cost I hadn't expected. I was asked to join and chair several committees, and this meant I was doing no research. Despite that, I did feel my job had purpose, and although I was extraordinarily busy, I was surprisingly happy. It also helped that Sonya, too, was very supportive.

Ben, too, was overloaded with administrative work that stemmed from delivering our vision for Biology. We have contrasting skill sets, and once detailed planning for the approved new building began, Ben took charge of ensuring it met department needs. I stood back from my role in fundraising for the building and became joint head of department, with Ben and I sharing the role. Eventually Ben stepped back as head of department and I took up the reins. I led the department through the pandemic, spending hours on Microsoft Teams each day, but by the beginning of 2021 the time had come round to vote for the next department head. Flatteringly, but frustratingly, the faculty voted for me to continue until the move into the new building. The problem I had was that I was no longer enjoying the job. We had delivered the vision, and that had given me purpose, but the day-to-day running of the department was not rewarding. It is a job that needs to be done, but I did not find it fulfilling. Nonetheless, I agreed to accept on the condition I could take a year's sabbatical in Australia. I needed a chance to resurrect my research and write this book.

Happiness is a strange thing, and it can sometimes be elusive, even though we often try to make choices to maximize it. One of the downsides of being as conscious as we are is that we are often aware that we are in pursuit of happiness. Yet sometimes we focus on the wrong goals. Keeping up with the Joneses and accruing vast wealth might not make you happy, but it is what we tend to focus on. Perhaps consciousness is not always as wonderful as we might think.

Plants and fungi are not conscious. They have no brain or equivalent organ to feel experiences of the world around them. They do not produce even rudimentary brainwaves. Although they are not sentient, they can still respond to environmental cues, growing towards the light and extending roots downwards, but these processes require no decisions to be made. Developments of particular parts of a plant or mushroom are towards a cue, be it light, gravity or water, that rudimentary sensing cells detect. These sensing cells do not link to neurons in a brain to be interpreted.

Unicellular organisms also lack the ability to be conscious, but they too can respond to environmental cues, using specialist organelles called cilia to move through their environment towards or away from useful and harmful chemical stimuli. These single-celled organisms can detect chemical cues in their environment, and as molecules that carry these cues bind to proteins on the surface of the cell, they can respond by moving towards or away from the cue.

Brains are the domain of animals, but not all animals have them. Adult sponges do not have brains, nor any form of nervous system. In some species of sponge, juvenile larvae do have simple brains. The larvae use these to move and to find somewhere to settle. Once a suitable piece of real estate is identified, the larva settles in a place to call home and begins to develop into an adult. As it does so, it digests its brain, as it is no longer useful. Several other sedentary animals, such as mussels and oysters, also lack brains. Those few species of animals that do not move and have a brain tend to have very simple ones. The ability to move seems to have been the evolutionary driver of developing a brain.

Why would the ability of multicellular animals to move drive the development of a brain? The evolution of muscles permits movement, which is the ability to, temporarily at least, resist the forces of gravity and electromagnetism. A rock, or indeed an adult sponge, is unable to resist these forces, but every time you walk, you use energy to resist gravity, lifting your feet and placing them in a new position. You also move through air, resisting the drag created by the electromagnetic force as you move. Being able to move opens up the way for choice and the option of making decisions: should I go over there, or stay here? The evolution of movement makes free will, and consciousness, a possibility, even if not a necessity. Complex mobile animals have choices, and evolution has produced complex brains, and presumably consciousness, to help organisms make the best possible choices. Swimming bacteria and archaea can move, and by this argument could have free will, but no within-cell brain organelle has ever evolved to allow decision-making.

Moving allows animals to escape from harmful environments, but to do this they need to sense harm. Even quite simple animals appear to be able to register pleasure and pain. Houseflies modify their behaviour after experiencing an injury, although it is unclear whether they feel pain in quite the same way as us. Nonetheless, if an animal has a nervous system such that they can transmit signatures from sensory organs to the brain and back to muscles, and act in response, it suggests that the sensation of pain is a characteristic that evolved at around the time the first mobile animals appeared on the tree of life. The first wisps of free will likely evolved with the ability to make directed movements. Pain is an effective way of telling an animal to act, and to do so quickly.

There is a terrible condition in humans called congenital analgesia that, fortunately, is extremely rare. Those who suffer from it cannot feel pain. The condition is caused by genetic mutations that prevent some types of neurons communicating effectively by impacting the flow of ions across the cell membranes that form synapses. Signals cannot be sent through the body's network of neurons in the usual way. Congenital analgesia is extremely dangerous, as individuals are frequently injuring themselves as they do not respond to harmful stimuli, and sufferers often die young from accidents. Registering pain is an important part of being an animal, and that is why fruit flies, shrimps, fish and snakes are all able to do it, even if they don't feel pain in exactly the same way as we do.

A key challenge in the science of consciousness is working out exactly what it is that different animals experience. How widespread is consciousness, and how does its degree differ between different species? In the last few decades, a growing number of scientists have concluded that even quite simple animals, including invertebrates such as shrimp, lobsters and octopus, can experience pain and probably other emotions similar to pleasure. They do not think that all these animals experience pain and pleasure in identical ways, and exactly what pain means to a crab, or a lamprey, is yet to be worked out. We do not yet know what it feels like to be a bat, rat or cat. Nonetheless, the ability to experience pain is a very ancient trait, it is probably the first emotion that any animal experienced, with higher degrees of consciousness, perhaps even complex thought, evolving from the ability of early animals to experience pain. When any individual experiences pain, regardless of the species, it will not be pleasant, and this means we do need to re-evaluate how we treat species like crayfish, lobsters, octopus, chickens, pigs and cows.

Scientists have compared the structures of brains in animals across the tree of life, and this has helped them work out how brains, and to a lesser extent consciousness, have evolved. The simplest brains consist of only a small number of neurons, with each looking the same. More complex brains, including ours, consist of lots of different types of neurons that are linked together in all sorts of different configurations. Different types of neurons in different areas of the brain can differ in shape and in the number of synapses they form, even though each neuron contains the same structures such as an axon and dendrites.

Brains evolved and became more complex by increasing the number of neurons, by diversifying their shapes, and by playing with the way they are connected. As we move through the tree of life from the first animals to you and me, we see that brains get more complex, with new areas being built on top of existing areas. Evolution has tended to add on new sections as brains have become more complex. Evolution is like a builder, adding on extensions when a new function provides a fitness advantage. Mammals developed the neocortex, something that is not found in other animals, and humans take the size of the neocortex to an extreme. During development, we grow lots and lots of neurons in the neocortex to a greater extent than most other mammals do. In contrast, Woofler developed a long nose with lots of cells that can detect smells along with a relatively larger part of the brain that enables him to interpret smells to a much greater extent than humans can. He can read the recent history of a patch of ground through smell in a way we can't, and this helps him locate prey to hunt. But evolution can shrink areas of the brain too. Some of our ancestors had better sight, hearing and smell than we do, but they could not solve crosswords or sudoku puzzles or master fire. On our branch of the tree of life, our brains have increased in size, but the relative size of different areas has fluctuated over the last hundreds of millions of years. We are at a point where it is beneficial to have a large neocortex, but perhaps as artificial intelligence becomes more human, we will lose the need for such an organ. Perhaps our descendants will return to the trees and lose the ability to speak while artificial intelligence does all the heavy lifting. Predicting the future is beyond this book, but understanding the past is not. We are now at a point in our history where mammals have evolved, and where animals are conscious, but I am yet to discuss how humans came to dominate the planet. That is the topic of the next chapter. Why are there humans?

Humans

At this point in our journey, mammals have evolved, and these mammals have brains that have given them a degree of sentience. These brains consist of neurons wired to form enormous networks that allow at least one mammal species – us – to understand and remember things about the world. The stage is now set to consider how humans came to be, and how they conquered the planet. We are an important species, but we are still just a species. One of many hundreds of millions to inhabit the Earth.

Every now and again a species – or group of species – evolves that changes the planet. Humans are one of these, but by no means the first, and they are unlikely to be the last. Species of Cyanobacteria changed the planet billions of years ago via oxygenating the atmosphere. The first eukaryote species heralded an era of complex organisms. The first predator created a way of life never seen before, and the first land plants opened up the ability of life to colonize continents. Humans are just another species in a list of those that have changed the planet. We are an unusual species in that we have altered the planet very quickly, but we are a product of nature just like every other. The evolutionary biologist Olivia Judson has a nice way of describing epochs of life based on the way organisms acquire their energy: 'geochemical energy, sunlight, oxygen, flesh and fire'. Humans are one of the first, but probably not the first, species to use fire as a form of energy. I focus on humans in this chapter because this is a book on why you and I exist. We are special because we have unusual attributes, but we are not so special we should think these traits mean we are in some way separate from, or above, nature. We are not.

The next step on our journey from the Big Bang to us involves understanding how modern humans, that have changed the world, came to be. You already know how evolution proceeds, so the next step is to consider how it moulded modern humans. Mammals needed to evolve from being small, solitary beasts into primates, and then into us. This required the evolution of several traits, and the ability to develop technology of increasing complexity, and that is the focus of this chapter. But before we start, let's consider the end point – modern humans.

One thing that we do as a species that is unique is create ways to make our day-to-day lives frustratingly difficult by putting in place layers of bureaucracy. You might think that because universities are full of supposedly clever people they might run smoothly. You'd be wrong. The University of Oxford where I work is administratively complex, but I am sure that any organization that has been around for over 800 years would have generated enough red tape to drive even the career bureaucrat round the bend. In October last year, a colleague's computer went missing during an office move. It was never found, and quite understandably she urgently wanted a replacement. She was told she could have one but would have to order it from a specific company. The company didn't have any in stock and had no idea when they might be able to supply it, but it would likely take a few months. A local business could supply the desired model for the same price within twenty-four hours, but my colleague was not permitted to buy from that business as it was not an approved supplier. The process to approve a new supplier can take several months, the outcome is not guaranteed, and as far as I can tell appears to be entirely random. Almost equally astonishing is that six different people have to approve any expenses claim I put in. I have no idea why this degree of scrutiny is necessary, but wouldn't be entirely surprised if it goes back to a ruling by Ralph of Maidstone, who was chancellor in 1231.

The head of the University of Oxford is the chancellor, which these days is a largely honorary and ceremonial role. His deputy is the vice-chancellor, and she is responsible for the day-to-day running of the university. She chairs a partially elected board of trustees that challenge or endorse programmes of reform. A group of pro-vice-chancellors report to the vice-chancellor, and those with portfolios are responsible for research, education, people, buildings or financials. Research and teaching are conducted within departments that group people by subjects such as History, Biology or Economics. Within these departments, various ranks of professors are responsible for delivering courses and conducting research, and they are managed by heads of department, such as me, who are ultimately responsible for hitting budgets, the successful delivery of taught courses, health and safety, and various other aspects of administration. Departments are grouped into four divisions called Humanities, Social Sciences, Medical Sciences and Mathematical, Physical and Life Sciences, and this level of organization is responsible for budget setting and strategy in these four broad areas. Thirty-nine independent colleges, some of which are extremely ancient and wealthy, admit students to courses, with many university-employed academics having second contracts with a college to deliver tutorials in support of the course the students they admit are studying. Staff employed by the university can join Congregation, which is the sovereign body of the university. Congregation can overturn decisions approved by the trustees, the vice-chancellor and the pro-vice-chancellors, but only occasionally does so. No other university is organized like Oxford, and although it has employed leading academics for over eight centuries, I am confident none of them would have advocated for the governance structure we have. A question that has long intrigued me is, how did evolution take us from a solitary mammal with a small brain to intelligent, highly sentient beings that work in complicated organizations and live in diverse societies that are so complex we do not really understand how they function?

Humans are the only species to have formed nation states and complex organizations with structures like the University of Oxford. We have been able to do this because of our abilities to think and communicate abstract ideas via language, coupled with most people's general acceptance of the laws of the nation state in which they reside, however crazy they may be. In this chapter I explore how this astonishing social and organizational complexity arose.

The first hominins, species in the same evolutionary family as humans, walked the planet about 6.5 million years ago, while the first cities were formed about 9,500 years ago, and it is the period between these dates I focus on in this chapter. Once humans had developed cities, art and writing, the rest is history (literally). There are very many good books on this topic, and I do not attempt to summarize human history over the last 9,000 years or so, instead focusing on the journey from the first hominin to the first city. A key role of science in piecing together what happened between the first hominin and the first city is in ageing bones, teeth and artefacts. Palaeontologists and archaeologists have pieced together this history from bones, artefacts and genetics, and although a new find can lead to a revision of this history, we now have a reasonable account of the journey from a fruit-eating primate to a space-age ape.

There are a number of key phenotypic traits that needed to evolve for humans to form complex organizations, cities and nation states. One is the ability to communicate with language. If we were unable to describe complicated ideas, explain why we were taking particular actions, or understand and articulate the motivation of others, complex societies could not have emerged. Woofler cannot learn what his nemesis dog next-door has been up to by talking to him, but I can find out via a conversation with Bilbo's owner.

The formation of societies also requires an ability to tolerate others and to live with unrelated neighbours. Coupled with this is our ability to form and accept hierarchies. If everyone felt a need to lead or be in charge, stable social structures could not evolve or persist. We are also able to distribute specialist tasks among members of a group of individuals in ways that are rare in the animal kingdom. It is not necessary for everyone to hunt large game, build houses, forage or cook. Each of us does not need to be capable of fulfilling all roles within society or an organization, but we can deploy our expertise while allowing others to do the same. Some redundancy is useful, as having two or three people who can hunt or cook in a group can provide backup if one is ill or injured. A third phenotypic trait is our ability to behave and make decisions now that may not be rewarded until some point in the future. We invest in stocks and shares to reap the possible returns at a later date.

Although these characteristics are key to our success, they don't always result in the same social structures, organizations or civilizations emerging. Humans are very flexible, and even in relatively small groups working together, different social structures emerge. The collaborative research group I have worked with on sheep and deer tended to be hierarchical, individuals tended to have quite strongly held opinions on the type of science that should be done, and there was little flexibility. In contrast, the group I have worked with on the guppies is much less hierarchical, more flexible and more open to new ideas. The wolf scene is much more political, in part because the presence of wolves in Montana, Idaho and Wyoming is contentious, and although the team are open to new ideas, there is always concern about political fallout. The sheep, guppy and wolf collaborative research groups have all been successful, and there is clearly no right or wrong way for scientific collaborations to function. Who knows, perhaps the different cultures arose because we simply end up resembling our study organisms? Sheep and deer form social hierarchies and can be pretty stubborn, guppies are less socially organized and highly adaptable, while wolves are highly social balancing within and between pack dynamics. I am sure I have changed as I have switched between study systems.

It is not just research collaborations that can culturally differ, so too do nation states. Different countries have different social, political and economic structures, with some being more open and others more dictatorial. Some societies are driven by religious doctrines, others are sectarian. There are many ways we organize ourselves, and many ways for societies and organizations to succeed. All this flexibility stems from our intelligence, the fact we are social, and our ability to think abstract thoughts.

Identifying at which points during human evolution these characteristics emerged is not entirely straightforward. Decision-making abilities, for example, do not fossilize. Even anatomical evidence can be a challenge to interpret: the evolution of an anatomical structure that allows vocalization of particular sounds does not provide evidence of language. Many birds, for example, are excellent mimics, with some species being able to repeat words and sentences with astonishing accuracy, but this is not an ability to communicate articulately in the way that we do. Palaeontologists have discovered from fossils that species that may well have been our ancestors had the anatomical ability to make a very wide range of noises at least 25 million years ago, yet they hypothesize that complex language evolved less than 2 million years ago. Related to this is the degree of language complexity needed to form a complex society. Are past, present and future tenses required, and what about conditional arguments? Apparently not, as some modern languages do not have all these tenses. I wonder if this book will be translated into one of those? Do individuals need a vocabulary of ten, a hundred, a thousand, or ten thousand words? It is presumably fewer than the average number of words we use today. The average English speaker knows about 40,000 words but has an active vocabulary of about half that number. Did inhabitants of the first cities 9,500 years ago know so many words?

Despite the challenges of working out when the key attributes required to live in complex societies first appeared in our ancestors, palaeontologists and anthropologists have made progress in identifying some key events, and some consensus has emerged. The first appearance of cave art, types of tools, jewellery and trinkets, evidence of deliberate burial, and use of fire all provide clues to the abilities of our ancestors. Other evidence is harder to interpret. *Homo erectus* was a hugely successful ancestral species of ours that dispersed from Africa into Europe and Asia. Populations of *Homo erectus* established on islands such as Flores and Java in Indonesia that could only be colonized by crossing tracts of ocean. Some scientists have argued this is evidence of seafaring abilities that would have required some form of language, while others contend that arrivals would more likely have been accidental, with individuals carried across on rafts of vegetation washed out to sea following ancient tsunamis. This latter perspective has *Homo erectus* as a species with an anatomy like ours but that lacked humanity. Without additional evidence, the debate cannot be confidently resolved, although I have an opinion that I will share later. What is agreed is that many skeletons from Flores were of a hobbit-sized race. Many species of large mammal including elephants, deer, rhinos and hippos evolve to smaller sizes on islands, and *Homo erectus* may have shown this pattern too, yet palaeontologists do not agree on this either, with some arguing the fossils are from individuals that may have been inbred, and that the diminutive size was due to developmental abnormalities rather than evolution. Evolutionary biologists also do not agree on what drives large species to evolve to smaller sizes on oceanic islands, but they hypothesize that a lack of predators, competitors or more benign island climates may play a role.

Telling the narrative of how humans evolved attributes enabling them to develop complex societies, and then how they then deployed them to do so, is difficult because the field of human evolution is fast-moving. Each new fossil find can lead to re-evaluation of significant aspects of the sequence of events. Two authors, David Graeber and David Wengrow, have recently produced a remarkably well-researched book, *The Dawn of Everything: A New History of Humanity*, which provides a major revision of previously accepted wisdom on humanity's roots. It has not been met with universal approval in the popular press or in the fields of archaeology and anthropology, but attempts to revise history rarely are. I found the arguments compelling. The debate the book's publication has sparked arises because there are relatively few data from our ancient past and interpreting them is like trying to describe the picture on a jigsaw puzzle with only a small fraction of pieces to hand.

The story I tell below is my interpretation of the current evidence. It is my conclusion that *Homo erectus*, one of our recent ancestors, was intelligent but not as intelligent as humans, but that Neanderthals and the first *Homo sapiens* had cognitive abilities very similar to ours. These latter species were not the dumb cavemen and -women so often portrayed in some popular books and documentaries. The rise of civilization took time because for it to happen, certain technologies needed to be invented, but such technologies only became necessary and useful at certain points in our history. There was no requirement to develop them before-hand. These points in our history were determined by food availability, climate and competition between different groups of hunter-gatherers. The story of the rise of complex socie-ties is a narrative of how technology our ancestors developed forced behavioural changes upon us. We are a product of our technologies, and likely will be into the future too.

Modern-day humans are a species of primate, a large group of mammals that includes the lemurs of Madagascar, monkeys and apes. All these species have eyes that face

Timeline of Human History

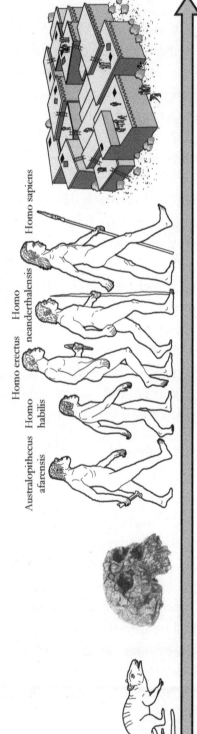

Australopithecus afarensis   Homo habilis   Homo erectus   Homo neanderthalensis   Homo sapiens

55 million years ago
– First primates

8 million years ago
– hominins and chimps diverge

3.9–2.9 million years ago
– Australopithecus afarensis

2.3–1.65 million years ago
– Homo habilis

2 million–110,000 years ago
– Homo erectus

430,000–40,000 years ago
– Homo neanderthalensis

300,000 years ago–present
– Homo sapiens

First City
c.9,500 years ago

forwards, and most species are agile and at home in the branches of trees. Primate hands are anatomically similar to their feet, and are highly dextrous. These hands and feet evolved to allow them to climb trees, handle and manipulate fruit and leaves in the canopy, and to catch insects, yet dexterous hands enabled our ancestors to develop the technologies described in this chapter. Good fossil evidence reveals the earliest primates lived in forests 55 million years ago, although some palaeontologists have made claims based on less convincing evidence that they date back to the Cretaceous, the geological epoch that started 145 million years ago and ended with the Chicxulub impact event that drove the dinosaurs extinct 66 million years ago.

Our closest living ancestors are chimpanzees and their close relatives, bonobos. The most recent common ancestor of chimpanzees and humans lived 5–10 million years ago, and may have survived until about 4.2 million years in the past. Fossil evidence of the common ancestor that everyone can agree on has not been found, with some paleontologists arguing that species of the European genus *Graecopithecus* is the most likely candidate while others favour the African species *Sahelanthropus tchadensis*. The disagreements will continue until new fossil remains of cousins of our ancestors are found.

What is clear is that our ancestors began to spend less time in trees and more time on the ground, living a lifestyle that was perhaps similar to that of modern-day baboons. The move to a more ground-based existence may have been due to a decline in forest cover in parts of Africa leading to a need to travel between trees as forests became savannas. As forests began to disappear as the climate changed, our ancestors' diets also changed, expanding to become more diverse. A more ground-based existence favoured upright walking, and this freed up dextrous hands for other tasks. Evolution took some time to deliver these changes, with them not happening all at once.

There is an emerging consensus that *Australopithecus anamensis* is the first species on our branch of the tree of life that is not on the chimpanzee branch. It lived in what is now Kenya and Ethiopia between 4.2 and 3.8 million years ago, by which point the human and chimpanzee branches had split. Male *Australopithecus anamensis* were about 1.5 metres tall, with females 20 centimetres shorter. The numerous fossils that have been discovered reveal that the species walked upright in a similar way to us, but with long arms that would have made them good tree climbers. They lived in a woodland habitat, still primarily eating fruit and vegetation, but they might have supplemented this diet with insects and small mammals and birds they were able to catch or scavenge. The species lived in small family groups, but although they were quite social there is no evidence of any tool use or culture, although some scientists have speculated that they may have used twigs and sticks to help forage in the way that modern-day chimpanzees do. The size of the skull suggests that species in the Australopithecine genus, including Australopithecus anamensis, were not as smart as their descendants, but they were possibly at least as smart as the non-human great apes alive today: the Bornean and Sumatran orang-utans, the Eastern and Western gorillas, chimpanzees and bonobos.

Several other Australopithecine species in addition to *A. anamensis* have been discovered, some from only single fossils. These fossils reveal that the Australopithecine lifestyle was successful, with fossils of descendant species found in central, eastern and southern Africa. The most famous fossil specimen comes from Ethiopia. It is a near-complete skeleton of a female, named Lucy, from the species called *Australopithecus afarensis*. The Beatles song 'Lucy in the Sky with Diamonds' was repeatedly played during the expedition that found her remains, and this was the cause of an individual of a long-dead species acquiring a modern-day name. Lucy lived at the peak of the Australopithecines' success, 3.2 million years ago, nearly 1 million years after the genus evolved and 1.8 million years before it disappeared. What caused their extinction is not known, but palaeontologists have speculated that either climate change or competition from a new species on the landscape that had evolved from an Australopithecine ancestor played a role.

When evolution discovers a successful new way of life, body shape or behaviour, individuals exhibiting these attributes increase in number, and the range of the species can increase as offspring move away from their parents in search of pastures new. The world is not homogeneous, however, and some individuals may end up living in environments that differ from those of their ancestors. Over time, lineages in different environments can start to diverge as they adapt to local conditions, particularly if there is no, or only very occasional, mating between individuals living in different environments. Evolution happens as populations in different areas adapt in different ways to the different environments in which they live. Australopithecines spread across much of Africa and in different environments they adapted to either more forest- or more savannah-dominated habitats. Their body shapes, diets and behaviours evolved to be different, and in time, separate species emerged. A mosaic of Australopithecine species living in a range of environments evolved, and somewhere, a species emerged that thrived in a drier, cooler, less forested environment than other species in the lineage. At the time, this species might have appeared a little peculiar, being the odd cousin in a family of otherwise largely similar-looking species, living in the margins that the other Australopithecines avoided.

As we have seen, eccentricities in the Earth's orbit, the gradual shifting of tectonic plates and volcanic activities can lead to changes in the climate. The Earth may appear stable and unchanging for centuries or millennia, but on longer timescales change is inevitable as the climate warms or cools and nature adapts. As Africa started to cool, prime Australopithecine real estate started to disappear, and as it did some species started to go extinct. But one species' loss can be another's gain, and the cooling world and the spread of savannahs favoured the odd cousin. In the fossil record we see the demise of the Australopithecines and the rise of the first species of *Homo*, the genus of closely related species to which we belong. If the fossil record were more complete, we might be able to link the first *Homo* species to the peculiar Australopithecine cousin, but currently scientists have not been able to do this. What we do know is that 2.3 million years ago a new species starts to appear in the fossil record, *Homo habilis*, which translates from the Latin to handyman. Nine hundred thousand years later the Australopithecines disappear from the fossil record, and the species of *Homo* that evolved from them were the dominant apes to walk the planet.

The gradual nature of evolution has meant that some authors have attempted to classify *Homo habilis* as an Australopithecine species, but currently it is considered to be the first species in the genus *Homo*, the species group that includes *Homo sapiens*. *Homo habilis* evolved from its Australopithecine ancestor to consume more meat, spend more time on the ground and less in the trees, to thrive in savannah environments, and in the use of stone tools. These tools were simply hewn rocks, formed by crashing rocks together to create smaller, sharper objects that were then used to butcher animals. The first human technology that became widely used was hand axes simply made from these knapped stones. These tools heralded the start of human-like species developing technologies that 2.3 million years later resulted in the LHC, a machine capable of probing the dynamics of the fundamental building blocks of matter.

Homo habilis are thought to have lived in larger groups than their ancestors, with some groups consisting of perhaps as many as eighty individuals, and this was a driver for increased intelligence. Scientists have identified a strong association between the size of the neocortex and the size of social groups within a species. The more individuals you have to interact with, the more brain power you need to manage the complex social interactions you are continuously faced with. Our neocortex size predicts we can cope with a social group of 150 individuals, which is about the average size of our social networks. This number is termed Dunbar's number, after Robin Dunbar, the British anthropologist who first estimated it. If *Homo habilis* did live in groups that were eighty strong, early homo species were already starting to become quite smart. Nonetheless, *Homo habilis*, although more human than the Australopithecines, still differed from us in several ways. Males were significantly larger than the females, and they cooperated to protect the group from predators and to scavenge meat from them. Although meat formed a greater proportion of the diet than in Australopithecines, examination of fossilized teeth shows that fruit and vegetation were key constituents of most meals.

Some descriptions of the lifestyle of *Homo habilis* draw comparisons with modern-day baboons and savannah chimpanzees. In all these species, males are considerably larger than females, with one or a small number of males monopolizing matings. Species where a small number of males conduct most of the matings are known as polygynous, and in such species males do not contribute to rearing offspring. It is usually assumed that *Homo habilis* was polygynous, although this is highly speculative as patterns of reproductive success across individuals cannot be fossilized. Perhaps females worked together to raise offspring, but some evidence suggests that *Homo habilis* offspring did not have a long childhood, being less dependent upon their parents than we are, and being independent at a much earlier age.

Regardless of the species' mating behaviour, *Homo habilis* was successful. Like the Australopithecines before, they, or species similar to them, appear to have spread to many parts of Africa, colonizing savannah habitats as the forests retreated. And just like their ancestors they too adapted to each environment they inhabited. Some of the skulls that have been found have an anatomical feature known as Broca's area (the motor speech area) that is essential for speech. Some scientists have interpreted this as evidence that some populations of *Homo habilis* might have developed rudimentary speech, although conclusively proving this is likely to be a challenge.

It seems unlikely that *Homo habilis* had all the attributes needed to develop nation states and complex organizations, but it did have some of them. Individuals could live in large social groups and cooperate with one another in a more advanced manner than their ancestors could. They may also have had a primitive form of speech, and they could produce simple tools, and such behaviour may suggest the first murmurings of abstract thought. Nonetheless, tool use was

rudimentary, and there is no evidence that *Homo habilis* produced art, used fire or buried their dead. Their descendants
did start to do these things, and next we turn to a hugely successful ancestor of ours, *Homo erectus*, or upright man.

If you were to see an individual of the species *Homo erectus*
walking in the distance, you would be excused for thinking
she was a human. Anatomically they were similar to us. They
were larger than *Homo habilis*, ranging in height from 145 to
185 centimetres, and weighed up to 70 kilograms. In addition
to their large size, they also had bigger brains than their ancestors, though these were smaller than ours. The average human
brain volume is 1,350 cubic centimetres, while that of *Homo
erectus* was a little under a thousand. This next step on the journey to you and me first appeared in the fossil record 2 million
years ago, 400,000 years before the last *Homo habilis* died.

Homo erectus differed from *Homo habilis* in that it was an
apex predator, meaning that, like us, it was top of the food
chain. *Homo* species had moved on from scavenging meat
from other carnivores to also killing their own. Groups of
the species hunted animals as large as wild cattle and elephants, as well as smaller prey, but they still supplemented
their meat diet with some fruit and vegetation. The Australopithecines and *Homo habilis* had been successful in colonizing
many environments in what is now sub-Saharan Africa (the
Sahara desert did not form until 11,000 years ago), but no
fossils of these species have been discovered outside of
Africa. In contrast, *Homo erectus* fossils have been found
throughout Asia and Europe.

Exactly how human *Homo erectus* was is hard to know.
There was variation in anatomy across populations, with
numerous other *Homo* species evolving from *Homo erectus*.
A bit like the Australopithecines before them, different

populations adapted to the different environments in which they found themselves. There are compelling arguments that *Homo erectus* colonized the island of Flores, before evolving into the diminutive *Homo floresiensis*, the hobbit human. Flores has always been an island, so its colonization required *Homo erectus* to cross the ocean. Whether this happened by chance, with a small number of individuals surviving the journey by clinging to logs or mats of vegetation, or by design in purpose-built dugout canoes or other forms of boat, is the subject of debate. The answer is important, because if *Homo erectus* was a seafaring species, it suggests an ability to communicate complex ideas and issue instructions of when to paddle in order to maintain a particular course. Unfortunately, the likelihood of a seafaring craft surviving for hundreds of thousands of years in the waters around Flores is extremely slim, so we may never know whether *Homo erectus* journeyed to the island by chance or by design.

Tool use was a ubiquitous feature across *Homo erectus* populations. The tools they produced were more complex than those of *Homo habilis*, with sharp edges chiselled on multiple sides, and stone knapping used to form sharp points. Hand axes of a variety of sizes were commonplace, with the tools used to cut up meat, wood and vegetation. The production of these tools was an important technological advance on the journey from the first primates to us.

Homo erectus roamed the Earth for nearly 2 million years, with the species disappearing from the fossil record about 110,000 years ago on the island of Java. Over that time, they evolved. In some populations they developed early forms of art by engraving shells, in others they tamed fire, and in some they built shelters. Given these abilities, it is not beyond the realm of possibility that individuals in some populations of

Homo erectus built seafaring craft. During the course of their 2 million years of existence, the *Homo erectus* brain would have evolved, and the neocortex likely grew in size, giving them more human cognitive abilities than their predecessors. Other human traits also evolved in *Homo erectus*, including monogamy.

Biologists can infer what the breeding system of an extinct mammal was by looking at the difference in sizes of adult males and females. In polygynous species, such as red deer, where a dominant male outcompetes other males before mating with many females and then plays no role in rearing offspring, males tend to be much larger than females. In contrast, in more monogamous species, where males and females form longer-term pair bonds and raise offspring together, males and females are of more similar size. In *Homo erectus*, males and females were similar heights, suggesting that the species was monogamous and hinting at both parents providing some parental care. If that was the case *Homo erectus* had evolved a different breeding behaviour from its *Homo habilis* ancestors. However, in some populations it appears that the *Homo erectus* infant brain did not develop much beyond birth, revealing that, as with *Homo habilis*, the young may have required less parental care than those of modern-day humans. Nonetheless, if they were monogamous, both parents would have likely provided parental care as such behaviour is seen in most other monogamous mammal species alive today, even if the length of childhood was relatively short.

Homo erectus lived in groups, but we know little about how big these were. Analysis of nearly a hundred fossilized footprints from a site in Kenya reveals that a group of twenty individuals walked through the site 1.5 million years ago. There are four sites where footprints are preserved, and one of the sets of footprints was made by a group of adult males.

Were they on their way to communally hunt? Some scientists think so, and they have interpreted this as evidence that different demographic groups may have carried out different tasks. *Homo erectus* may have been the first hominin where individuals, or groups of individuals, regularly performed specialist tasks beyond defending the group from predators.

As we saw with the Australopithecines and *Homo habilis*, success leads to diversification, and the same was true for *Homo erectus*. The species produced many descendant forms beyond *Homo floresiensis*, including the Neanderthals, Denisovans and *Homo sapiens*. *Homo erectus* conquered Africa, Europe and Asia, and evolution sculpted each population to the environments it encountered. The Neanderthals evolved to live in the forested and often cold climes of Europe, the Denisovans inhabited Asia, while modern-day humans arose in Africa. Fossil remnants from Denisovans are sparse, so we know little about their lives, but Neanderthals and modern-day humans shared many similarities. They both developed art, language, complex societies and they both buried their dead. Both species pushed the cognitive abilities of *Homo erectus* to new levels.

In the text above, I have simplified the story. There is a line of descent from *Australopithecus anamensis* through *Homo habilis* and *Homo erectus* to *Homo sapiens*, but there were many offshoots. Populations would have diverged from one another, with their descendants sometimes crossing paths and successfully reproducing. We know this happened between modern-day humans and Neanderthals when *Homo sapiens* left Africa, and more on that later. Evolution is rarely the straightforward progression the tree of life suggests. Populations can diverge, and then hybridize when they come back together, or individuals from different populations can mate, and if their offspring then preferentially mate with each other, new species can evolve. Both the Australopithecines and *Homo erectus* would have done all these things, and our genome is consequently a mix of genes that evolved in different places. The fossil record is a testament to this, with a wide variety of hominin forms found across the globe. Palaeontologists have described several species or subspecies of *Homo*, including *H. rudolfensis, H. ergaster, H. georgicus, H. antecessor, H. cepranensis, H. rhodesiensis, H. neanderthalensis, H. floresiensis* and *H. heidelbergensis*. Palaeontologists cannot agree on which of these should be classified as species, subspecies, or diverged populations. I tend to think of them as subspecies. Regardless, the number of forms shows how remarkably successful *Homo erectus* was at adapting to new environments.

One intriguing form is *Homo naledi*, which lived in South Africa between 335,000 and 235,000 years ago. The dating of fifteen individuals to the relatively recent past is surprising given the anatomy of the species. Individuals were small, averaging about 140 centimetres in height and weighing 40 kilos. They had relatively small brains for their body size, but the shape of their skulls suggests a modern brain structure. The bones that have been recovered were from a single site in the Rising Star cave system in Gauteng province. What is remarkable is that the bones of no other animals were found in the chamber where the *Homo naledi* fossils were discovered, and access to the cave would not have been straightforward a quarter of a million years ago. The bones were not washed into the cave, but instead the corpses of the dead may have been placed there, suggesting deliberate burial. *Homo naledi* exhibits a mix of Australopithecine and *Homo erectus* characteristics, yet they were alive at the same time as modern humans. Exactly how this unusual species is related to other early hominins remains to be discovered.

What is clear is that many different types of hominins lived across Africa. Biologists refer to cases when many different closely related subspecies live close by as a radiation. Hominins appear to have quite easily adapted to different habitats within Africa and beyond, with species like *Homo habilis* and *Homo erectus* being particularly adaptable. As adaptation occurred, over the course of about 4 million years, our ancestors evolved many of the key characteristics that make us human and allow us to live in complex societies.

The above narrative briefly summarizes the development of early hominins, but it does not explain why they evolved the characteristics that led to us. We are unlikely to ever know for certain, but as already mentioned, some of the Australopithecines began to spend less time in the trees and more time on the ground as their forest homes disappeared as the climate became increasingly dry. The species would have evaded ground-based predators by taking to the trees, but some began to spend more time in more open savannah habitats. The more successful lineages developed strategies and behaviours to protect their groups from the big cats, bears and wolves that shared these environments. These strategies may have involved throwing stones, wielding sticks, and working together to scare off or even kill predators. Bigger individuals may have been favoured as their strength and speed may well have given them a survival advantage. As anti-predation behaviours evolved, an unexpected benefit was the ability to steal kills from other predators and eventually hunt them. This was made easier by the development of more complex tools and through social cooperation in joint hunting. Those individuals that were most able to effectively work together and to build the most useful tools were able to outcompete groups that were less well equipped or organized. The adoption of fire and the construction of shelters would have been adaptations that further aided survival.

These adaptations required a larger brain, and brain size has trended to become larger over evolutionary time in hominins, with the odd exception of species such as *Homo naledi* that appear to have briefly thrived with a relatively small brain. A large brain has many advantages, enabling improved communication, problem-solving abilities and abstract thought, but big brains are expensive to run, and difficult to give birth to. If you are an average-sized human, your brain will account for about 2 per cent of your body weight but will use 20 per cent of your energy. Despite the cost, having a large brain was clearly worthwhile for our ancestors.

At least two hominin lineages, Neanderthals and modern-day humans, evolved large brains, and both appear to have mastered fire, cooked, worn clothes, developed language, produced art, made jewellery, constructed musical instruments, used medicinal plants and cared for the sick and injured. Neanderthals lived in the forested areas of Eurasia between 430,000 and 40,000 years ago and were well adapted to the cold winters they experienced. They were stockier than modern-day humans, a characteristic that may have helped them retain body heat during winter, but which may also have enabled them to sprint quickly when hunting. They were top of the food chain, able to successfully hunt mammoths, bison and other large herbivores, but they also ate smaller animals, plants, fruits and nuts. At least some individuals had larger brains than modern-day humans, but we too may have smaller brains than the first humans. Perhaps we are not as smart as once we were, or, alternatively, our brains may have become more efficient.

Neanderthals are usually thought of as primitive, thuggish cavemen who were easily outsmarted by our ancestors. Careful examination of Neanderthal sites reveals a different story, and as palaeontologists have studied Neanderthal tools, jewellery made from shells, a flute, the ash of hearths in caves and their anatomy, it has become clear that intelligent apes have evolved more than once. Neanderthals lived at low densities, and their total population size may never have exceeded a few tens of thousands of individuals, but such low densities are not unusual for top carnivores. Meat was an important component of their diet, and the availability of prey could have prevented them from achieving larger population sizes, as carnivorous species tend to live at lower population densities than those that eat more vegetation. Despite their low densities, at times different groups met and likely traded objects, food, innovations and ideas. The creation of art, jewellery and music reveals what psychologists called symbolic thought, the ability to use words and pictures to represent objects or events that are not currently happening. Neanderthals were not stupid.

As well as being able to think in an abstract manner, Neanderthals also modified their environment and may have kept gardens. In a site near Leipzig, Germany, archaeologists have uncovered over 20 hectares of land that were kept deliberately open by Neanderthals for over 2,000 years approximately 125,000 years ago. It is not clear whether the site was deliberately cleared by Neanderthals, or if a natural event led to its clearance, but the vegetation record includes species that only grow in open areas, and the site is archaeologically important, with the area littered with evidence of Neanderthal activity spanning two millennia. The surrounding area was forested, and if the Neanderthals had not kept it open it would have reverted to forest in a few decades. We do not know why they kept it open, but it was a deliberate action that continued for twenty centuries, so the reason must have been important to them.

Neanderthals also produced the first known cave paintings, beating modern-day humans by about 20,000 years. The site of these 65,000-year-old paintings is near Santander in north Spain, and the art consists of black and red handprints and hand stencils, depictions of animals, and various other lines and shapes that are harder to interpret. By way of comparison, the oldest known *Homo sapiens* art is a life-sized painting of a pig painted 45,000 years ago in Indonesia.

Throughout the course of humanity's history, we have tended to consider ourselves as the centre of everything and to be more developed than other life forms. Humans used to assume that the Earth was the centre of the universe, and everything revolved around it. That proved to be wrong, as the Earth orbits the sun, so we made the sun the centre of the universe, but that too was wrong. We now know we are on a planet orbiting an average star, towards the edge of an average galaxy, that may contain as many as a trillion planets. Despite this, we often still see ourselves as the pinnacle of evolution, with no other species having evolved language or intelligence, either on Earth or perhaps anywhere else in the universe. It turns out that this bastion of our uniqueness is also crumbling. Two types of intelligent ape evolved on our home planet, although they did share a common ancestor in *Homo erectus*.

If Neanderthals were so smart, why did they go extinct? Their extinction occurs shortly after *Homo sapiens* settled in Europe, so it seems we might take some responsibility. But as is often the case, events are a little more complex. *Homo sapiens* first left Africa sometime between 177,000 and 210,000 years ago, with fossils from this period unearthed in Greece and Israel. They did not thrive in Europe, though, and subsequently died out, possibly being unable to compete with the local Neanderthal populations. A second, more successful dispersal attempt out of Africa occurred 60,000 to 70,000 years ago, with people crossing into what is now the Middle East and working their way around the coast of Asia to Oceania. These *Homo sapiens* made quick progress, with the first modern humans crossing into Australia at least 40,000 years ago, and perhaps as early as 60,000 years ago, but *Homo sapiens* were not to successfully spread through Europe until 40,000 years ago. Did they not know it was there? Did they not like the look of it? Or did attempts made between 40,000 and 60,000 years ago to move into Europe fail?

The fossil record that palaeontologists have discovered cannot answer these questions. My suspicion is that something prevented humans from colonizing Europe earlier, and that it may have been that Neanderthals were better adapted for life in Europe than *Homo sapiens*. Those early humans that tried were not able to thrive given the competition they met from the resident Neanderthals, encountering similar resistance to the first wave of the *Homo sapiens* that left Africa. However, eventually something changed, and *Homo sapiens* were able to make progress into Europe, and that something was the climate.

Homo sapiens were well adapted to the hotter, but sometimes drier, environments of Africa, while Neanderthals were specialized to live in the forests of Europe. *Homo sapiens* were the intelligent tropical ape, and Neanderthals were their smart, temperate cousins. Changing climate between 45,000 and 40,000 years ago brought a drier spell to Europe, and this altered the environment, tipping the balance in favour of humans over Neanderthals. As humans spread through Europe, Neanderthals died out. The exact cause is unknown, with arguments made for disease, competition from *Homo sapiens*, decreasing populations of the large game Neanderthals relied upon, and even inbreeding. Although the last Neanderthal died forty millennia ago, their genes live on. As our ancestors spread through Europe they interbred with Neanderthals, and these genes may have provided some advantage, as they have remained within human populations to this day.

The fact that Neanderthals and modern humans interbred reveals they viewed members of the other group as potential mates. We do not know whether matings were consensual or forced, or whether humans and Neanderthals formed pair bonds. They would not have spoken the same language, having evolved apart for thousands of years, but given they both had the ability to learn languages it is not beyond the realm of possibility that they jointly raised young. Unless we find remains of a blended family we may never know, but what is clear is that children born of Neanderthal–*Homo sapiens* matings became parents themselves. If they didn't, there would be no Neanderthal genes to be found in the genomes of modern-day people. The human genomics company 23 & Me reports that I am more Neanderthal than over 80 per cent of people who have used the company to be genetically profiled. I'd love to know how I'd have turned out without those genes.

It was not just in Europe that modern humans mated with local hominins. As humans spread through Asia, they mated with Denisovans. Genes from this mysterious hominin are present in several Asian and Oceanic populations of modern-day humans, but fossil evidence of the group is scarce. Nearly everything we know comes from genetic studies, including the one described earlier in the book. The fossils that have been discovered reveal that Denisovan hominins were like Neanderthals, but with slightly different teeth and perhaps other anatomical differences too.

Following their successful dispersal out of Africa, and their colonization of Asia, Oceania and Europe, humans crossed a land bridge 26,000 years ago between what is now Russia and Alaska, going on to spread throughout the Americas. By 4,000 years ago, humans were established in Polynesia, and, except for Antarctica, could call land from the tropics to the Arctic their home. But had climate change gone in the other direction many millennia ago, would the Neanderthals have dispersed into Africa, displacing *Homo sapiens*? We will never know, but in an alternative universe, perhaps it is the humans who are portrayed as the dumb cavemen, while it would be the Neanderthal sending rocket ships into space. That wasn't to be, and this book is about humans, so what happened next? We are now at a point in our story where we are the only hominins left on the planet, and we went on to build civilizations.

When *Homo sapiens* left Africa they would have had similar intellectual and reasoning capabilities to you and me. It was a slow road from the upright intelligent ape with limited technology to the world we live in today. How did it happen? How did evolution create a modern man from a clever caveman? The first buildings on the site at Çatalhöyük were constructed about 9,500 years ago in what is now Anatolia, southern Turkey. By 9,000 years ago it had become the first city. It persisted for nearly one and half millennia, and at its peak had a population of 10,000 people. A lot had to happen in the more than thirty millennia between our ancestors leaving Africa and settling in Çatalhöyük. The story now switches from biological evolution to the development of technology. By 40,000 years ago our ancestors had our cognitive abilities, but what they did not have is the technology. We now start the journey from a stone hand axe that led to a smartphone, a story in which our dextrous hands and our large brains proved key.

Following dispersal from Africa, our hunter-gatherer forebears expanded into a wide range of environments and fed on creatures as diverse as elephants, elephant seals and the now extinct elephant bird of Madagascar. Numerous hunter-gatherer cultures arose, with evidence from extant tribes suggesting that many may have shared resources and other possessions equitably within the group, and that women and men had equal rights. Over time, these groups developed and adopted novel technologies that helped them more effectively acquire resources from the environment, with successful technologies spreading across the ancient world as they were copied by one group from another. Our desire to keep up with the neighbours is not a modern vice.

In the modern world technology tends to equate with electronics, and the invention and widespread adoption, in western societies at least, of dishwashers, smart fridges and computers has had an enormous impact on society in recent decades. This is not the first time that novel technology has altered human societies. The lives of our hunter-gatherer forebears were improved by advances in wood, stone, bone and metal tools, through the production of twine, string and rope, the innovation of pottery, through the ability to start and control fire, the construction of dwellings, and via the advent of cooking. These days we take these advances for granted, but each required keen observation, impressive abstract thought, and trial and error. Imagine being a member of the first group to be able to make fire. Your neighbours would view you with awe, assuming, that is, that your experiments in producing flames had not left you badly burned.

Fishing is another example of an important technology. The analysis of bones from a 40,000-year-old human from Eastern Asia revealed he frequently ate freshwater fish, although it is unknown how they were caught. The hunter-gatherer societies of the Andaman and Nicobar Islands are accomplished at fishing with spears and bows and arrows, and both technologies have been around for a very long time, so it seems likely our ancestors would have employed these techniques to fish. The earliest wooden spears date to 400,000 years ago and were produced by early Neanderthals or their *Homo erectus* ancestors. The spears are the earliest examples of wooden tools and although the ones that have been found were used for killing large animals such as horses, the same principle could easily be applied to fish.

The first evidence of a bow and arrow is more recent, dating to between 60,000 and 72,000 years ago, from Sibudu cave in South Africa. The technology spread quickly, and our ancestors who left Africa took bows and arrows with them. We know this because arrowheads have been found in Fa Hien cave in Sri Lanka next to fossilized remains of *Homo sapiens*. The technology was desirable because bows and arrows allowed people to injure and kill animals at a greater distance than could be achieved with spears. Using them was less risky. We do not know when bows and arrows were first used to kill fish, but as with spears, following their invention their use in fishing likely followed quickly.

Various other fishing innovations followed. The first

evidence of poison applied to arrowheads dates back 24,000 years, and it could have been used to stun or kill fish in lakes or rivers, although finding definitive evidence of this will be likely impossible. The oldest fishing net dates from 8,300 years ago, was made from willow and was discovered in Finland in 1913, although the discovery of stone fishing-net sinkers from Korea suggest nets were in use 27,000 years ago. Intriguingly, the invention of harpoons – spears with barbed bone hooks attached – appears more recently, being recorded in cave paintings from France just over 6,000 years ago. Their use quickly spread across the ancient world, and they were used to kill seals, other sea mammals and fish.

The exact dates of these discoveries are likely to change, with each new finding pushing back the date of the earliest use of a particular type of technology. We are often surprised at the great age of quite complex technology, but we should not be, given our ancestors were as clever as we are, and perhaps brighter. Timelines for the development of pottery for storing water and food, for the use of other types of stone, wood and bone tools, and for the use of fire have been documented, and they all tell the same story. *Homo sapiens*, along with Neanderthals and possibly other descendants of *Homo erectus* such as Denisovans, were intelligent and innovative, and their ingenuity allowed our ancestors to spread and thrive in environments from the Arctic to the tropics, from deserts to forests, and from low-level oceanic islands to the Tibetan plateau, an inhospitable landscape many days journey from the coast.

In numerous different environments our ancestors became more accomplished at gaining resources, both via improvements in their ability to hunt and in the breadth of resources they consumed. Many prehistoric cultures learned to grind up and soak, ferment or boil unpalatable foods to remove compounds that were hard to digest. Acorn flour is an example. The seeds of oaks are nutritious, being rich in oils and starches, yet they are bitter to eat and hard to digest due to a class of chemicals called tannins. Acorns are more nutritious, and tastier, if the tannins are removed, yet this is not a straightforward task.

To remove tannins from acorns they need to be ground up and soaked. Online instructions for making acorn flour on websites providing hunter-gatherer recipe suggestions involve grinding the acorns before soaking them for up to a week, changing the water daily. The acorns are best dried with the shell and skin taken off before being ground up. The tannin removal process can be sped up by boiling. The resulting flour is tasty as well as nutritious and was consumed by hunter-gatherer groups whenever oaks were abundant. Humans have been making flour from acorns and other nuts and seeds for over 30,000 years, although the earliest bread so far found dates back only 10,000 years.

Making flour not only removed tannins, it also protected our teeth. The human bite is not particularly strong, and trying to chew on hard food can damage teeth. In the absence of a dentist, tooth decay and loss can be a serious issue, reducing the amount and types of food that can be consumed. Grinding seeds, nuts and animal bones was consequently an important innovation, and once it was invented those groups that used it would have had a survival advantage over those that did not. The technology consequently spread, and as it did so, local populations that adopted it would leave more descendants.

Any population has a replacement rate, which is the average number of offspring each individual must produce for the population to remain the same size from one year to the next. If you and your partner have two children, and both survive until adulthood, you will have replaced yourselves. If those two children each find partners and themselves have two children that survive to adulthood, they too will have replaced themselves, and you and your partner will have four grown-up grandchildren. When this happens to everyone, a population will remain the same size into the future. Yet not all individuals survive from birth to adulthood, and this means that the replacement rate in human populations with good healthcare provision is a little over two children. In cases where mortality between birth and sexual maturity is higher, the replacement rate is also higher, and the hunter-gatherer replacement rate may have required each woman to produce between four and six children. However, they did better than this as hunter-gatherer populations spread to conquer the globe, and this means their reproductive rate was higher than the replacement rate.

A population that does not continually increase or decline in size has some process that keeps its size within bounds, and the process that does this is described as 'limiting'. Populations of wild animals are often limited by predation, food availability, disease, suitable breeding sites, or even the ability to find a mate. Early in our history, *Homo sapiens* may have spent a 70,000-year period soon after they evolved from *Homo erectus* being restricted to a small patch of green, lush land in Botswana in an area that is now Lake Makgadikgadi. The population would have been stable for millennia and expanded only following climate change. Not all scientists accept this interpretation of the data, but it would not be surprising if our ancestors' population was limited by predation, food availability or disease at some point in its past.

Technological advances freed our ancestors to reproduce at a rate faster than the replacement rate. An increase in the ability to acquire resources, and to reduce the rate of mortality from predators, meant that local populations would have increased. As the local population increased, the amount of food available on the landscape would have declined. As life became harder, individuals began to disperse, and it was this process that led to humans covering most of the globe. Each technological advance allowed our ancestors to acquire more food and to produce more young. But more mouths require more food, and this would have driven down available resources. Without the advent of newer technology, some people would have upped sticks and moved to pastures new.

Humans are heterotrophs, which means they get their energy from feeding on other lifeforms. The success of an individual heterotroph comes at a cost to something else. Grazers thrive by eating grass, browsers by eating trees and shrubs, and carnivores by eating other animals. Humans are no different, and it is why we are negatively impacting our planet's natural environments. You might think that this is a recent innovation, but it is not. We have impacted nature since our earliest days, and we will continue to do so into the future.

Humans are bad at sharing their lands with other large animals unless they have domesticated them. Across the globe, people persecute wolves, lions and bears, and we are in the throes of driving rhinos and elephants extinct. Even large marine organisms, such as whales, seals, and large sharks that pose very little risk to us, are not spared. Sometimes we persecute them because we fear them, other times it is to eat them, to turn them into trinkets, or to satisfy the misguided belief that some attribute of an animal's organ will manifest in you. Our modern-day dislike of large animals is not a new phenomenon. As humans colonized new parts of the globe, they routinely drove large species extinct. Some scholars shy away from blaming our ancestors for the carnage because changing climate also often correlates with the demise of some large animals and this means it can be argued to play a role. The issue with blaming the weather is, first, that climate change allowed humans to conquer new lands, and second, many of the disappearing species had survived earlier periods of climate change without dying out. We have never been good at living with large animals, and quite probably never will.

A second line of evidence for our negative impact on nature comes by looking at the distribution of animal body sizes on continents in the presence and absence of our ancestors. An examination of the body sizes of animals on each continent 125,000 years ago suggests that early *Homo sapiens* rid landscapes of large animals from very early in our history.

The distribution of body sizes found on a land mass is related to its area, such that larger bits of land tend to have higher average body weights of animals. Africa is one of the world's largest continents so we would have expected it to have many large animals, but 125,000 years ago, by the time *Homo sapiens* had evolved, that was no longer the case. It contained fewer large animals than predicted by its size. The big beasts were already largely gone. As humans moved across the world, they continued the killing. Species such as the elephant bird of Madagascar, the giant ground sloth of South America, the marsupial lion of Australia, the giant cave bears of Northern Eurasia, and woolly mammoths were all driven extinct. Their demise was our ancestors' success, but eventually there was nowhere new to disperse to, and new technology was required for our populations to grow.

As local hunter-gatherer populations increased in size, neighbouring groups would encounter one another more frequently, opening the opportunity for trade, and war. In some parts of the world, different hunter-gatherer groups from different environments began to seasonally meet up to trade goods and perhaps find partners. Structures began to appear, such as Earth mounds and early buildings, in places like Lepenski Vir in Serbia, which hunter-gatherers used as a seasonal meeting place. The next step was towards living year-round in one place, and the development of living in the same house year-round came hot on the heels of another key innovative technology: agriculture.

The oldest evidence of agriculture comes from the shores of Lake Galilee, 23,000 years ago, where edible grasses were cultivated on a small scale by the Ohalo II people. Despite the obvious ability to farm crops, widespread adoption of agriculture in the region took a further 11,500 years, by which time pulses such as lentils, peas and chickpeas, and grasses including wheat, rye and barley, were cultivated. Climate change may have contributed to the declines in animal abundances, along with hunting, and such change could likely have also altered the distribution of nut and fruit trees in the region, leading to agriculture becoming a necessity.

Although agricultural innovations were copied in similar ways to other technologies, it was independently developed at least eleven times across the globe. Different cultures grew different crops, with rice being domesticated in China, cassava in the Amazon basin, and beans and squash in the mountains of the Andes. It was not just plants that were farmed for food. Pigs were domesticated 11,000 years ago with sheep and cattle following shortly afterwards. Chickens were domesticated about 3,500 years ago in South-east Asia, although some scientists suspect it may have occurred as many as 4,500 years earlier.

Other animals were domesticated to aid with transportation or moving large objects, or for other purposes. Horses were domesticated on the Steppes of Central Asia 5,500 years ago. Dogs were domesticated at least twice, perhaps as long ago as 35,000 years ago, although this date is not agreed upon by all researchers. Fish farming is 3,000 years old, and emerged in China.

All this reveals that our ancestors understood that plants and animals could be grown for food, and this was a useful way to supplement foraging from the wild. However, agriculture was time intensive, hard work and not always conducive to the nomadic lifestyle hunter-gatherers enjoyed. Agriculture also required an excess of seeds – in times of food shortages a better use of seeds would be their consumption rather than sowing them. Early agriculture may not have been a necessity, but perhaps a backup plan in case of future food shortages. Evidence of this comes from opportunistic planting of crops in silt deposited on floodplains following receding floodwaters. Relying on such events avoided the need for clearing land. Crops were then left and returned to later, rather than being continually tended and protected from wild animals.

As the amount of technology increased and humans got better at hunting, fishing, growing crops, removing unpalatable compounds from food, building shelters, making clothes, looking after domesticated animals, healing the sick, and creating pottery, jewellery and other items, their numbers increased, but so too did the inability of one person to master all the technology required, or to maintain or carry all the tools required. Experts in particular technologies would have arisen, particularly when their use produced surpluses of food. Trade began, and humans became more sedentary, in part because specialists in the use of some technologies benefited if they could be easily found. Villages first appeared at the beginning of the Neolithic period, 10,000 years ago, as specialization increased.

Coupled with the rise of the specialist, as human population size continued to climb, our ancestors became more sedentary. As food security became less certain due to declining animal populations and climate change, our ancestors began increasingly relying upon agriculture. This further tied them to particular locations, and the size of some villages and towns began to increase.

For reasons that we shall soon come to, archaeologists believe that life in the first city, Çatalhöyük, was egalitarian. The city was divided into two halves by a gully, which may have separated two or more separate peoples who chose to live as neighbours. Anthropologists have argued that Çatalhöyük lay on the boundaries of three distinct ecosystems, each of which contained hunter-gatherers relying on different resources. The site of Çatalhöyük may have consequently started out as a meeting point between these groups before people started living there permanently as trade and specialization became more important. The city consisted of private houses, each with three or four rooms, entered via a hole in the ceiling that doubled as an exit vent for smoke from cooking fires. Much activity appears to have happened on the roofs, including the production of artefacts, trinkets and statuettes of figures that strongly suggest religious beliefs. No public buildings or temples have been found at Çatalhöyük, and all the houses are similar in construction, which is why archaeologists believe the society was egalitarian. Hunted and gathered food was a significant constituent of the diet of the residents, with Çatal-höyük appearing to be a hunter-gatherer society writ large, further suggesting that the adoption of agriculture did not drive the development of permanent cities and towns.

Although societies in some early towns such as Çatal-höyük were egalitarian, many later towns were home to more hierarchical societies. Many of these settlements were constructed on high ground, or were in bends of a river, or in places that could be approached only from one side, making them easier to defend. Raids by neighbouring people, or non-settled nomadic peoples, were clearly a threat, and ways of minimizing these risks were factors in deciding where our ancestors chose to live.

Evidence of thriving hierarchical societies comes from some of the earliest cities in Mesopotamia. Large, ornate and imposing temples were at the heart of each city. Such edifices reveal that organized religions were an important part of life in these cities. Early religious beliefs were animist and based on life spirits. Life-after-death beliefs were widespread, with ancestors often believed to return in animal or mystical spirit form. The dead were often helped on their way to the afterlife, being buried, sometimes with belongings to aid their journey. Although shamans and other religious figures were believed to have the ability to commune with the spirits and were respected, they did not accumulate significant wealth nor use their sacred abilities to impose hierarchies on societies. This was to change as villages, towns and cities emerged, with the construction of temples and more organized worship, and the wider uptake of philosophies in which one's destination in the next life was a result of one's behaviour in this. Elites based on religion emerged, with the early Mesopotamian cities having a religious king surrounded by a ruling elite. Several of these kings went on to construct narratives of their derring-do, providing them with the status of god-king. The religious hierarchies that arose, coupled with expertise in some technologies being seen as more highly prized than expertise in others, led to hierarchical structures and top-down control emerging. Rules of acceptable social behaviour and punishable antisocial behaviour were established, and the populations of cities were compelled to follow these behaviours in return for salvation in after-lives and increased security in this one. Most people would accept these arrangements if life was not too hard, but when the elite became too tyrannical, or life became hard because the elite failed to appease the gods, they could be overthrown. Security required a fighting force that could be used to maintain order in times of peace. Those who disliked arrangements could leave, or if they too vociferously opposed the elites they could be banished, imprisoned, enslaved or killed.

The maintenance of a hierarchy always needs funding, and in return for the taxes that were imposed the elites built and maintained temples and other public buildings and facilities. Dissent and disobedience occurred if taxes were too high for the facilities provided, yet many monarchs and chiefs were able to accumulate vast wealth.

This structure of many societies has remained largely intact since. It appears to be a stable arrangement for human organization. Advances in technology have allowed city size and human population to continue to exponentially increase, and now there are 8 billion of us living on the planet, with some cities being home to more than 20 million individuals.

The degradation of the environment the earliest humans started has continued and has now led to us changing global carbon and nitrogen cycles and the composition of our planet's atmosphere. For example, the availability of nitrogen in the form of compounds that plants can use constrained the productivity of most ecosystems up until the 1950s. This all changed when Fritz Haber and Carl Bosch invented a process to manufacture nitrogen-based fertilizers on an industrial scale. Since then, humanity has liberally used such fertilizers to increase crop yields, with over 200 million tonnes sprayed on to crop in recent years. As rain has washed some of this fertilizer from agricultural lands into rivers, lakes and oceans, humanity has altered the functioning of ecosystems not only on land but also in fresh and salt water environments. In a related manner, at the time of the first civilization in Mesopotamia, carbon dioxide concentrations were at 260 parts per million compared to over 400 parts per million today. Our love of burning fossil fuels is changing the atmosphere and along with it the weather.

Some people are rather dismissive of the impact that we are having on our environment as our civilization feels secure. We are not the first people to think like this. In his excellent book *The Earth Transformed: An Untold History*, Peter Frankopan catalogues correlations between past changes in the climate and other aspects of the environment on the rise and falls of civilizations past. It doesn't take much to shatter illusions of stability. When supply chains required to maintain large populations break down, individuals struggle to survive and the key structures that are so central to existence disintegrate. When ancient civilizations have collapsed, they have done so quickly, and often with great loss of life. Many ancient and long-lived civilizations collapsed, with few signs of them remaining today. If history teaches us any lesson, it is to be thankful for periods of environmental stability, as these have been good for humanity. We change our climate and the functioning of global ecosystems at our peril.

The people who lived in the first cities were anatomically the same as us and had the same intellectual capacity. The history of human civilizations is fascinating, and science has played a key role in piecing this history together by being able to non-invasively scan ancient archaeological sites and artefacts, and to use tools such as radiocarbon dating to estimate the age of objects. Yet the stories of these civilizations and the people that lived in them, although fascinating, are not central to an understanding of our existence. Different civilizations uncovered different truths of our universe as new technology was produced, largely by trial and error. The Greek and Roman empires may not have had to happen for you and me to exist, even if they did demonstrate new ways for societies to organize themselves. Civilizations have come and gone, with the key thing about humans being that we are capable of building them. The invention of science needed civilizations, and without them we could not have pieced together so much of the narrative of our existence. Details of the rise and fall of human civilizations have been told many times before and are beyond the focus of this book so they only get a brief mention.

Civilizations that have collapsed include the Mycenaeans, the Hittites, the Western Roman Empire, the Mayans, the Mauryan and Gupta civilizations of India, the Angkor of Cambodia, and the Han and Tang dynasties of China. These collapses were often triggered by climate change and resource shortages that resulted in civil unrest and war, and appear to have often taken most inhabitants by surprise.

Societies, and complex organizations within them, are fragile. A combination of an overly authoritarian leader with a natural catastrophe can lead to the collapse of empires. Similarly, on occasion, more empathetic leaders have united people, enabling them to survive periods of hardship. Why do leaders differ in their approach to leading? What is it that causes individuals to be different? That is the focus of the next chapter. Why are we the way we are?

You and Me

I have so far told the narrative of how the universe, the Milky Way, our Solar System and the Earth came into being. I have explored physics, chemistry and biology, the emergence and spread of life, the evolution of consciousness and the rise of humanity. I have covered 13.77 billion years of the past, and we are nearing the present day and the end of this extraordinary journey. Then I can answer the question I posed at its beginning: were we inevitable at the birth of the universe, or are you and I just incredibly lucky? As the history has unfolded, things have become ever more localized as we have moved from the history of the entire universe to the rise of the first civilizations in a small corner of our planet. The last step is more local still, and it involves exploring why you and I are as we are.

I face two challenges in navigating this part of the journey. First, I do not know you. One reason I have described a little about my history is so you have an idea about the sort of person I am. I focus on me because I cannot answer questions as to why you might be short or tall, optimistic or pessimistic, shy or outgoing, why you might be scared to snorkel but happy to jump out of a plane, or why you might have remained single or are on your fifth marriage. Second, it is challenging to identify the cause of particular characteristics within one person. At the level of the individual, science can struggle to explain why you are anxious or relaxed. Answering questions about individuals requires understanding the detailed genetic and developmental mechanisms that have made you an introvert or an extrovert, shy or bombastic, and currently we do not fully understand these mechanisms. Nonetheless, scientists have made progress. They know that our genes, our experiences and chance all contribute to our characters, and the focus of this chapter is on these three processes.

Because I do not know you, I have put myself and occasionally my wife Sonya under the spotlight throughout this book. Before I dive into the topic of this chapter, I will briefly summarize key aspects of my journey. My childhood was happy with supportive parents, but I didn't enjoy school much. Despite this, there was much to read and satisfy my curiosity at home, which meant I did a lot of learning outside school. A spell in Africa as a young man changed the course of my life and inspired a desire to understand why I exist. Because I had little idea of what to do for a living, I did a Ph.D., and followed an academic career trajectory; I still study the ecology and evolution of animals. The job involves publishing academic papers, although between 2017 and 2024 I took on leadership roles with the University of Oxford, which means I have sacrificed my research to focus on building a legacy for current and future biologists at Oxford. Sonya, my children and Woofler are very dear to me and are a constant source of happiness. This is my first book, and I wrote it without much of a plan on what to do once the first draft was finished, but I found a wonderful agent and publisher, and that is why you are reading it now. The reason I have told this personal story will now hopefully start to become clear: I want you to know some of these details to illustrate how experience, genes and chance can influence us all.

Lots of measurements can be taken from individuals such as adult height, birth weight, how extrovert we are, anxiety levels, sprinting speed at age twenty and intelligence. Your birthweight might be six pounds five ounces, and your sprinting speed at age twenty might be 15.6 mph, and these are called phenotypic trait values.

Values of any phenotypic trait vary across individuals within a population. You were probably a faster runner than I was at age twenty, but you are likely to be shorter than I am. I can say this with confidence because I am below average in running ability, but I am taller than the global average height. I am five foot eleven, while the global average for men is somewhere between five foot seven and five foot nine.

If the values of phenotypic traits such as running speed at a particular age, the strength of the immune response you mount to fight an infection, or anxiety level are plotted against how frequently they occur within the population, a bell-shaped curve often describes the data well. A curve of this shape is sometimes referred to as a normal or Gaussian distribution. Carl Friedrich Gauss was the mathematician who first described this curve, and it is often named after him. The reason for the bell shape is that there are few individuals with either very small or very large values of a trait of interest, with most individuals falling in the middle of the distribution. If we take running speed as our x axis, with faster runners appearing to the left of the Gaussian distribution and slower ones to the right, Usain Bolt will be far to the left-hand tail of the bell shape because he was an exceptionally fast runner as a young man. I would be further to the right, beyond the highest point of the bell. The highest point of the curve is the average, or mean, value, while the width of the bell measured at a particular point on the graph is called the variance, and it quantifies the degree to which individuals in the population differ from one another. Scientists then identify factors that might explain some of these differences between individuals, and for phenotypic trait values these factors can be split into three classes: differences in our genes (nature), differences in the environment we experience (nurture), and something called developmental noise that is currently unpredictable and is thought to be due to chance.

The statistical approach of explaining variation in phenotypic trait values allows scientists to make statements such as 'smoking shortens lives'. There is irrefutable evidence that this is true because on average within a population life-long smokers do not live as long as those who have never smoked. However, this does not mean that smoking will lead to an early death for everyone who consumes a packet of twenty cigarettes each day. The average smoker dies younger than the average non-smoker, but we cannot accurately predict at what age any individual will die, regardless if they smoke or not. Some smokers live to old ages, avoiding the smoking-related illnesses that can kill many other smokers, and some non-smokers die young because they develop lung cancer despite never having dragged on a cigarette, cigar or pipe. Scientists can make statements about what happens on average within a group of individuals, but they cannot say with certainty what will happen to an individual. In the same way, we can say that traumatic events early in life on average result in introspective adults, but we cannot state that a child who experiences such an event will become introspective. They are just more likely to do so. It is frequently impossible to say with complete confidence that the reason a person has a particular value of a phenotypic trait is down to nurture, nature or chance. It is consequently impossible to state definitively why each of us is the way we are, but we can identify probable causes and roles for genes and the environment.

I have already described genes and summarized how they work, but it is worth briefly recapping. A gene is made up of long strings of molecules called nucleobases of which there are four types, A, T, C and G. Reading your genetic code would rapidly get monotonous as it is a very long string of these four letters. Nucleobases are grouped into threes, with each group of three instructing the molecular machinery in your cells to grab a particular type of amino acid – a type of molecule made from carbon, oxygen and nitrogen atoms arranged in a particular configuration that enables them to form chains.

Imagine the first three nucleobases in a gene are CCC. The triplet tells the molecular machinery to grab an amino acid called proline from the molecules in your cells. The next triplet in the gene reads GCA, which codes for the amino acid alanine. The molecular machinery selects an alanine molecule and joins it to the proline molecule, with this process continuing until a stop sequence such as TGA, TAA or TAG is encountered which tells the molecular machinery that the protein's production is complete. The molecular machinery that translates the genetic code to a protein contains lots of steps involving many types of molecules that I do not describe here.

The chain of amino acids then folds itself to create a protein. Proteins run our metabolism and are used to build our bodies. Something as simple as a change in a single amino acid in the chain can alter how effectively a protein does its job. For example, if the nucleotide sequence of a gene is changed even just a little, the sequence of amino acids in the chain may change, and the effectiveness of the protein produced may be impacted. If a nucleotide triplet CCC were to change to CAC, a proline in the chain would be replaced with a histidine and the chain might no longer fold as it should. When a mutation happens that changes the sequence of amino acids in the chain, the protein might stop working altogether, its function may be slowed, and in occasional cases it may even work more efficiently. When the order of amino acids in a chain is altered due to a change in the sequence of nucleobases, a genetic mutation has occurred. Each of us has different mutations – different variants of the human genetic code – and these differences can contribute to differences between us in phenotypic traits such as how outgoing we are, our birth weight, adult height, or our ability to run 100 metres quickly at age twenty.

I know of one genetic mutation I have because it prevents a protein functioning properly. I have a mutation in a gene called GPR143 that produces a protein with the snappy name of G-protein coupled receptor 143. The protein consists of 404 amino acids, and in my case at least one of these is not what it should be due to a mutation. I do not know how the mutation I have alters the order of amino acids in the protein the gene codes for, but the protein does not function as it should. In the database of human genes and what they do, GPR143 is described as encoding for 'a protein that binds to heterotrimeric G proteins and is targeted to melanosomes in pigment cells. This protein is thought to be involved in intracellular signal transduction mechanisms. Mutations in this gene cause ocular albinism type 1, also referred to as Nettleship–Falls type ocular albinism, a severe visual disorder.' In more accessible language and drawing on descriptions of how the eye develops, this means that when the protein works correctly it acts to add colour to a small group of cells in the back of the eye called the fovea. My fovea lacks colouration, leading to a genetic condition known as ocular albinism. Such a small effect – a few uncoloured cells – might sound trivial, but it does mean my eyes wobble, and this is the cause of my poor vision. My eyes do this because the colour that most people have in the fovea helps stabilize their vision, and my vision is not stabilized as it should be. My mutation means I have about 25 per cent of normal vision, and this means I can only read down to the third row of the eye charts at the opticians. My eyes have always wobbled and always will. It hasn't had too many detrimental impacts on my life other than that my eyesight is not good enough to allow me to drive safely. I must live within cycling or walking distance of work, and I am always the designated drinker if Sonya drives us to a social event.

The mutation I have means there is an error in my genetic code. Your genetic code is a self-assembly manual on how to build you, and the human genome is the generic self-assembly manual on how to build a person. My instruction manual to assemble a human eye has an error in it. Most of the instructions are correct, but the bit that says 'now add colour to these cells' is missing. My copy has a typo in gene GPR143, and this means my eyes did not develop as they should. To understand what this means I turn to the science of developmental biology, the science of describing how we grow and how organs such as our heart, kidneys and eyes are formed.

Complex multicellular organisms such as you and I start out as a single cell that then divides. Such cell divisions keep occurring until a small ball of a few tens of identical cells exists. From that point onwards, different cells start to develop in different ways. Those on the outside of the ball will go on to develop into skin, while those in the centre will become our internal organs. The way that cells 'know' how to develop is due to chemical signals. The concentrations of these chemicals determine when to turn particular genes on or off, and this determines how each cell develops and whether it will become a nerve cell, a muscle cell, a neuron in the brain or a blood cell in the veins.

As development continues, cells start to move around within the developing embryo, migrating to particular areas. In the brain, for example, the end of a neuron is guided into its final position, where it joins with other cells by means of things called neuronal growth cones, that are a little bit like molecular tractors that move along concentrations of particular chemicals in the developing brain. These chemicals act like signposts, telling the molecular tractor to keep going, head right or left a little, or stop.

By the time we are fully grown, our body consists of over 30 trillion cells. Many of these cells migrated into their final positions within the developing body following gradients of chemicals. These chemicals are produced by genes, or by reactions involving the proteins that the genes code for. Your genome does not contain a map of where each cell in your body will be and how they link to one another. Instead, it consists of a set of instructions for when each protein should be produced in each cell type, and when the rate of production of that protein should be turned up or down or turned off. These on and off instructions result in gradients of chemicals in your developing body, and these chemicals allow your body to construct itself. Your genome codes for instructions that state 'turn on if such-and-such a chemical is below a certain concentration, and turn off otherwise'.

Development is complex, and this means that most phenotypic traits are not determined by a single gene. Your adult height is determined by how long and how quickly your bones grew. Tall individuals have produced longer bones than those that are smaller because their bone cells continued to divide for longer during development or they divided more quickly. Growing bones is a complex task, with genes controlling the rate at which they develop, the directions in which they grow, the shapes they develop into, and the length of time they grow for. Your height is consequently not determined by one gene but by very many that control all these aspects of bone growth. There are many genetic ways to be tall or short, thin or fat, funny or serious.

When scientists discovered genes they did not have this understanding of how they control development. Early in the genetic revolution, when scientists discovered they could read DNA sequences, some biologists thought they would find single genes for nearly all our phenotypic traits. A gene for intelligence, another for life expectancy, and a third for the ability to play cricket. Some researchers dreamed of using genetic engineering to make people smarter, longer-lived, or better batters at cricket or baseball. Many benefits could be seen if genes were to work like this, with all sorts of diseases potentially becoming treatable. Perhaps a murderous psychopath could be turned into a thoroughly decent chap following some genetic engineering. But it was not to be. As we have learned more about the way our genes work and our bodies develop, we now see our genomes as self-assembly manuals, explaining how to build a human from a single, fertilized cell. There is no genetic homunculus determining your exact developmental trajectory from conception. The values of most phenotypic traits are controlled by lots of genes, and my genetic condition, ocular albinism, is unusual in that the extreme phenotypic trait value I have for my eyesight can be attributed to a single gene. Because only one gene causes ocular albinism, it could be a candidate for gene therapy. Despite this, my mutation does not operate completely independently, and although all males who have the mutation develop poor eyesight, we are not all equally partially sighted.

There are many members of my extended family who carry the mutation in the GPR143 gene and several men who have poor eyesight because of it. Although we all have worse than average eyesight, some of us can see better than others. The mutation in the GPR143 we have means we all lack pigmentation in our fovea and that impacts our sight, but other genes involved in the development of the eye mean that the impacts of the mutation on our eyesight differ between us. These other genes contain instructions on which types of cells to produce and where they should be positioned. No gene works independently when building a body or a phenotypic trait such as adult height, and this means that identifying how a particular gene influences the value of a particular trait is usually very hard. The reason that biologists know GPR143 influences vision is that all males who carry my mutation have eyesight that is a long way from the average of the bell curve, and geneticists have worked out what the protein the gene produces does. Biologists have also worked out a lot about how the eye works and have shown how important colouration in the fovea is for normal eyesight.

The self-assembly instruction manual for a human body is so complex, and involves so many steps and chemical gradients, it is perhaps not surprising that if you run the identical code multiple times it does not produce exactly the same result. Variations in chemical concentrations that we cannot predict caused by molecules bouncing around off one another within and between your cells as you develop mean that if multiple bodies are produced by the same genome, they are not completely identical but can have different phenotypic trait values. Biologists refer to these unpredictable impacts on development as developmental noise, and they have produced a number of clever methods to explore the impact of chance on determining our phenotypic traits. For organs that come in pairs like the kidney or feet, biologists can compare how these differ within individuals. Many of us have better eyesight in one eye than the other, with my left eye being slightly less bad than my right. Each eye develops in the same way, using the same genetic code and developmental steps, so any differences are due to events that we cannot predict and that consequently appear to be random. Biologists can also use differences in phenotypic traits between genetically identical twins to investigate the role of developmental noise.

Identical twins have identical genomes and they often look very similar, but there are differences between them that their parents are usually very quick to spot. These differences are due to developmental noise caused by small differences in the concentration gradients that developed within each twin as they grew. Such developmental variation means that if you were to create a hundred clones of yourself, they would not be perfectly identical. They will each have developed in slightly different ways, with each having a slightly different height, or face shape, or brain structure. Some traits seem to be more susceptible to developmental noise than others, with some psychologists claiming that some bits of personality may be strongly impacted by developmental noise. A hundred identical clones of you might vary in their degree of extroversion, with their individual scores producing a bell curve, with one or two being extrovert, one or two being shy and the rest being somewhere in between.

I have described how two processes, genetic differences

between us and chance, can influence who we are, but there is a third process to consider, the environment. The term 'environment' captures a lot. It includes, among other things, the country where you were born, the month and year of your birth, the culture you experienced while growing up, the altitude at which you spent your youth, your socioeconomic class, the amount and types of food you ate as a child, the people you grew up surrounded by, and events you have experienced. Identifying which aspects of the environment to measure and to focus on in scientific research is hard.

In most scientific studies the environment is characterized quite coarsely, using variables like class, the religion of the household you grew up in, the number of siblings you have, whether you are the oldest or youngest sibling or somewhere in the middle, nutrition during development, or the cohort of individuals you spent your formative years with. A typical study will correlate values of a phenotypic trait with aspects of the environment such as these. As an example, many studies have found birth order correlates with aspects of personality. In this case, some measure of personality such as anxiety level is the phenotypic trait, and birth order captures an aspect of your environment as a child as it correlates with the amount of attention your parents were able to give you. The firstborn child tends to be conscientious and cautious, while the youngest is more fun-loving, attention-seeking and manipulative. Those in the middle are more rebellious and have a large circle of friends. These differences are thought to arise because the oldest, being the first, gets lots of parental attention. Each middle child receives a little less, with the youngest having to compete the most for attention against its older siblings that still require parental care. Although these effects are undoubtedly real, you shouldn't read too much into them. Some firstborn children are rebellious and manipulative, and some last-born offspring are conscientious and cautious. Your birth order means you are slightly more likely to have a particular personality trait rather than that you will. Birth order explains only a small part of why people differ, with scientists saying it explains only a small proportion of variation in personality between people.

Although a single characteristic of the environment such as amount of parental attention explains only a fraction of variation in phenotypic trait values such as anxiety level, this does not mean the environment is unimportant. Instead, what such results reveal is that no one aspect of the environment, such as birth order, explains many of the differences between us. When scientists include many measures of the environment into analyses, such as birth order, the type of school you went to, how many overseas holidays you went on with your parents, or the ages of your parents, the total impact of the environment on phenotypic trait values increases. The environment is devilishly complicated, it is hard to quantify, but for many phenotypic traits it plays a key role in how they develop.

The genetic code is long and complex, development is complicated and involves many steps, and the environment is a multifaceted web of lots of different things. As if these drivers of phenotypic differences are not complicated enough, things become even worse when the influence of some aspect of the environment on phenotypic trait values depends upon your genes. Biologists refer to these effects as gene-by-environment interactions.

Smoking is something that is bad for you, and exposing your lungs to cigarette smoke is part of the environment you experience. On average, smokers do not live as long as non-smokers, in part because they are more likely to get various types of cancer, including bladder cancer, as well as the more obvious lung cancer. A smoker is at least three times more likely to develop bladder cancer than a non-smoker. However, not all smokers face the same risk of developing bladder cancer, with the chance of a smoker getting it being dependent upon the alleles they have at a gene called NAT2. Smokers with some genetic mutations at NAT2 are about three times more likely to develop bladder cancer than those with a normal copy of the gene. Your genetic code at gene NAT2 influences the likelihood of poisons in cigarette smoke causing a mutation in a cell in your bladder that makes it cancerous. Your genetic code is said to interact with an aspect of the environment, influencing the chances of developing a nasty disease.

Much of my focus in this chapter so far has been on physical traits such as eyesight or running speed at a particular age, but we also have personality traits. When we describe ourselves, we often mention things such as whether we tend to get stressed, are laid back, or have a fear of spiders. These aspects of our personality are also phenotypic traits, and they too can be influenced by genes, the environment and chance. We are often intrigued by why one sibling is shy and another outgoing, or why we are anxious but our partners are not. The reason is down to nature, nurture and chance, with nurture a function of the environments you have experienced. Despite the complexity of personality traits, psychologists have made progress in exploring the role of genes, chance and the environment on our personalities.

Personality traits come from the structure of our brains and the way brain cells link to one another via synapses. I discussed synapses and how they are involved in the ways our brains work when I explored consciousness. Brains are extremely complex, consisting of between 80 and 100 billion neurons linked by upwards of 600 trillion synapses. It is these linkages that determine how we experience the world, what we remember, how we think and why we have the personalities we do. Psychologists who study the brain do not have a mechanistic understanding of why the networks created by synapses between brain cells make some of us extrovert and others introvert, make some people scared of spiders and others indifferent towards them, or indeed any aspect of our characters. Despite this, psychologists have been able to work out which parts of the brain are associated with particular personality traits, and that genes, chance and the environment all impact our personalities much in the same way that they determine our physical phenotypic traits.

Some of the first breakthroughs in understanding brain structure were achieved by studying how unfortunate accidents have damaged parts of the brain in some patients and how these injuries have altered their personalities. Scientific papers on these cases typically refer to patients by numbers to protect their true identities. In one paper, scientists report how a thirty-year-old man referred to as Patient 2410 suffered a brain bleed towards the front of his brain during surgery. Before the injury his wife described him as short-tempered, often angry, and mopey. The surgery changed him. Afterwards he was much more laid back and spent more time laughing and joking. Another example of a positive personality change was seen in Patient 3534. At age seventy parts of the front of her brain were damaged when a tumour was removed. Her husband, who had been married to her for decades, described her as being stern, irritable and grumpy pre-surgery. Post-surgery she was happier, outgoing and chatty. Sadly, brain injuries do not always result in increases in happiness. They often turn once happy people into more troubled and aggressive beings.

Brain-imaging studies, where patients lie motionless in large machines which create 3D models of their brain, have allowed psychologists to identify parts of the brain that are active when a person thinks about a particular topic or event. Such work has revealed that groups of neurons in different parts of the brain are linked and are associated with positive and negative feelings, anxiety, despair and happiness. These scans have also revealed there are many differences between each of our brains, with no two brains being the same. We are each wired a little differently, with this wiring due to difference in our genes, our experiences and developmental noise.

Our brains, and to some extent our personalities, change with age, and this is because the way our brains work is not constant, as networks of neurons can be repurposed. Many of us will know children who were shy and reserved but who developed into outgoing adults. We may also know outgoing individuals who have become more introverted following a traumatic event. These changes occur because the role of neuronal networks in your brain is a little different from year to year. Some synapses linking neurons are reinforced by experience, while others fade if they are rarely used, with others being completely repurposed. Although we cannot predict how a particular network of neurons will result in a personality trait, we do know our brains are flexible, and that who we are changes from one month to the next as connections in our brains change as we experience the world around us.

These experiences are a function of the environments we encounter. As previously mentioned, for other physical phenotypic traits, environment captures a lot, including nutrition

and socio-economic class, but also things that made us nervous, scared, happy or excited in the past. And because we
learn, many of the environments that impact the wiring of
neurons within our brains are not random. As we develop,
we learn what makes us ecstatic, and what makes us fearful
or uncomfortable, and we tend to select environments we
feel comfortable in. If you are scared of heights, you won't
become a mountaineer. I don't enjoy dance clubs and so
actively avoid them. My experience of dance clubs is consequently limited, and they have contributed little to who I am.

Personality traits, like any other phenotypic trait, are determined by genes, chance and the environment, but they are
harder to study because we do not understand the mechanistic cause of traits like introversion or extroversion. Unlike
my poor eyesight, where scientists understand the genetic
error and how it manifests, we do not have a good grasp of
which brain structures equate with outgoingness or humour.
Without a mechanistic understanding of how brain structure
maps to personality we must rely on statistical correlations,
but this too is not straightforward, as my next example
demonstrates.

Sonya has a fear of sharks and is never comfortable going
into the ocean. She tells me that her fear started after seeing
two of her friends swimming in the shallows and spotting a
large shark swimming towards them. She shouted and waved
a warning and they safely swam ashore, with the shark turning back to the depths. There is no doubt that the experience
was frightening, and for a few seconds she was not sure how
it would play out. Nonetheless, if that event hadn't happened
would she be unfazed by the idea of swimming in the
ocean? Was she already a little nervous about entering the sea
before this event? We can never know with certainty because

a version of Sonya who did not witness the shark close encounter does not exist, and so we cannot make such a comparison. Because each of us is unique, we cannot do the experiment where we compare versions of ourselves that did and did not experience a particular event.

Even if my wife had an identical twin sister who had not experienced the shark close encounter and who was happy to swim in the ocean, I still could not say for certain that the day on the beach when her friends could have been injured or killed was the cause of Sonya's fear. Some other differences in the experiences of identical twins, or even apparent randomness during development in the womb or early childhood could be the cause of their different views of the safety of the sea.

Although scientists cannot attribute cause to effect within an individual in terms of why they have the personality they do, they can draw insights by comparing populations of people. If I were to identify 1,000 individuals who had observed their friends experience a close encounter with a shark, and 1,000 who had not, I could measure their levels of anxiety when I take them to the ocean and ask them to swim. I might do this by measuring changes in levels of adrenaline or cortisol, hormones associated with stress, in their blood. If the two groups of individuals did not differ in other ways, I might be able to conclude that seeing a shark swim towards your friends was sufficient to induce a fear of entering the ocean. I would reach this conclusion if individuals in the close-encounter group were, on average, more reluctant to enter the water than those in the non-close-encounter group. To draw this conclusion, the two groups must be as well matched as possible for all aspects of experience, biology and background, other than having seen their friends encounter a shark while swimming. If everyone in the close-encounter group was female, for example, while the non-close-encounter group was all male, I could not distinguish between sex and shark-close-encounter experience in determining fear of the ocean. Perhaps men are just more likely to be scared of the ocean than women and this could explain the difference between the two groups.

Even if my two groups were indistinguishable from one another in all other aspects of their lives except for the shark-close-encounter experience, I could not say with complete confidence that any single individual fears the ocean because she experienced a shark encounter unless all those who experienced a near shark attack were scared of the ocean, and all those who did not were happy to dive into the sea. In most comparisons like this, you do not find that all of one group exhibit one view, while all the second state another. It would be highly likely that some of the thousand individuals who had not experienced a shark close encounter were still scared of the ocean, and vice versa. The best I can say is that, out of a balanced and random sample of 2,000 people, those who experienced a shark close encounter were more likely to fear the ocean than those who had not. I can't say they will, or will not, fear the ocean, just that they are more likely to do so. Sonya did experience a shark close encounter so it is possible that this has caused, or contributed to, her fear of the ocean, but neither she nor I can ever really know despite her vociferous insistence that this was the cause.

Despite scientists' inability to link particular experiences to personality traits, all of us have experiences that we believe have moulded our personalities. Some people recount events to explain why they are shy, anxious, funny, talkative, or scared of almost any object you can imagine. For example, a woman from the West of England cannot walk down the frozen food aisle of a supermarket because of a fear of frozen peas. She explains, 'They tend to just look at me – ganging up on me.' She wasn't involved in a terrifying pea prank as a youngster, or a collision with a lorry carrying a shipment of peas, but maybe something triggered it when she was younger. Regardless, the fear is very real and it is part of her personality. With something as complicated as a personality, how can psychologists make progress in working out why we have the characters we do?

The first step is to simplify the problem, and psychologists frequently do this by classifying people on five dimensions of personality: extroversion, neuroticism (or emotional stability), agreeableness, conscientiousness and openness to experience. These are called the big five personality traits, and are part of a framework developed in the 1990s to assign individuals into categories by having them complete a questionnaire. Psychologists like these traits because they capture some of the differences between individuals, and scores for any one individual tend to be very similar if the same individual does the test on multiple occasions spanning months or even years. If acquaintances are asked to complete the survey for someone, there also tends to be agreement between the classifications provided by the subject and her colleagues.

By analysing brain scans, aspects of the environments in which people have lived and genetic sequences of study participants, psychologists have been able to show that, as with many other phenotypic traits, genes, the environment and developmental noise contribute to determining the big five personality traits. Environmental drivers such as nutrition as a child, socio-economic background or the country you are

Big Five Personality Traits

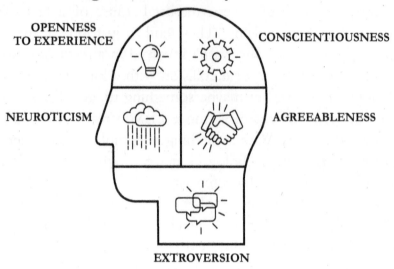

OPENNESS TO EXPERIENCE

CONSCIENTIOUSNESS

NEUROTICISM

AGREEABLENESS

EXTROVERSION

born in tend to explain little of the differences between us. You might think this means that the environment is not important in determining our personality, but that would be wrong. It would be a strange world if all people born rich were shy and all those born poor were not, or if the French were much more anxious than the Germans. The influence of the environment is important, but proving it is hard because individual experiences mould our personalities. Extremely traumatic events, such as abuse as a child, have been shown to impact openness to experience. Fortunately, such traumatic abuse is unusual and most of us do not experience such extreme environments, but that is not to say less traumatic events do not shape who we are. Sonya is convinced her fear of the ocean stems from one experience that lasted less than a minute, and certainly before it she had a job that required her to spend a lot of time in the ocean studying turtles. It is possible she had subconscious anxiety

about the ocean that only manifested itself consciously following her close-encounter experience. But scientists cannot
prove that.

Despite the enormous difficulties of studying personality
traits, we are confident they are like any other phenotypic
trait in that genes, the environment and chance determine
their values. They are harder to understand because classifying the environment in a way suited to studying personality
traits is hard and because a single experience might change
your personality but it wouldn't change your eye colour or
height. Although I know I can't prove it, I believe that my
genes and a few key events really have given me my character. Telling the story of these events is useful as it allows me
to turn to the role of personal narratives for making sense of
the world, and of who we are.

I enjoy life, and most of the time I do not find it too stressful. When I conducted an online survey to score myself on
the big five personality traits, I came out as more extrovert
than most of the population, emotionally reasonably stable,
more agreeable than four out of five folk, at least averagely
conscientious, and open to new experiences. I have been told
I am intelligent, but those who have said this have never seen
me attempt to put together flat-pack furniture.

I have not always had these personality traits. As a child I
rarely felt confident, was not particularly popular with my
peers, and often found social situations stressful. I was not
conscientious with respect to schoolwork and would rather
read books and magazines or watch TV, but for as long as I
can remember I was fascinated by science. How did I get
from the geeky, lazy and shy youth to who I am now? It is
due to my genes, the environment I have experienced and
apparent chance. I also believe, irrationally, there are a few

experiences that changed my personality and my outlook on life.

The mutation at gene GPR143 that impacts my eyesight changed the course of my life when I was a child. My father was a successful pharmacist before he retired at sixty, and this meant that my parents were reasonably financially secure when I was growing up. Dad had gone to the Perse School for Boys in Cambridge, the premier private school in the city, and he thought it would be a good option for me. As I neared my eleventh birthday and the time to move from primary to secondary school approached, my parents arranged to meet with the headmaster of the Perse school. The meeting initially went well, and the head of the school was encouraging that they apply for me to attend and that I would be likely to be offered a place. Towards the end of the meeting my father mentioned my genetic condition and poor eyesight, and explained I would not be opening the batting for the school's cricket team. On reflection, the headmaster stated, it turned out I probably wasn't a great fit to the school. He apparently wanted his students to be competent cricketers. Instead, I went to Comberton Village College, a state school in a local village. I suspect that this school was a better choice of school for me, and that the headmaster of the Perse school was correct, but for the wrong reason. My mutation caused a change in my educational environment, the friends I would go on to make and the teachers who struggled to educate me.

I was lazy at secondary school as I didn't enjoy the drudgery of learning. I liked finding out facts for myself, via experimentation, but I didn't enjoy learning lists of facts. My dislike of rote learning meant I was hopeless at French, German and Latin, because I refused to learn vocabulary. Because most of the schooling involved rote learning, I did not engage with it. The subjects I liked were maths and science. I liked science because I could do experiments. I devised a school project with my dad where I grew broad bean plants in different types and amounts of soil. Each plant got the same amount of light and water, and I would measure them daily to monitor their progress. It was fun, as I was finding out stuff for myself, and this revealed there were rules to the world. I enjoyed discovering these rules, rather than lists of facts such as the principal cities and towns of all the counties of the British Isles.

Maths was a little like science. There were rules on how numbers could combine or equations could be manipulated. If I understood the rules, I wouldn't need to learn my times tables but could instead easily work out the answers from first principles. From an early age, I wanted to know why the world worked as it did, rather than learning enough facts to help me navigate it. But it wasn't all plain sailing. I was taught calculus at secondary school by being given a set of recipes to integrate and differentiate various mathematical functions. I didn't understand where these recipes came from, and that frustrated me. Up until this point I had been good at maths and was in a class slated to do an examination certificate half-way through the two-year course. I achieved 6 per cent in the mock exams, possibly the lowest mark in the class. On the night before the exam, the penny eventually dropped as I thought things through, and as I stared at the recipes I suddenly understood where they came from and what differentiation and integration were doing, and I could see why they were such incredibly useful tools. The next day I did the exam. I hadn't learned the recipes but I understood how the bits of calculus I needed worked, and I came fourth highest in the class, being one of only a handful of students who passed. I wished we had been taught why calculus worked, rather than simply how to apply it.

At the age of seventeen, with my A-level exams looming, I was supposed to decide what I wanted to do with my life. Training in mathematics would surely be a safe bet for a reasonable job and so I applied to university to study it. I subsequently received offers, subject to securing the right A-level grades. At around the same time I was applying for university we had a presentation from an ex-pupil who had taken a gap year. He had spent the year overseas teaching in a school in Honduras. I enjoyed travel, this sounded fun, so I decided to apply to Project Trust, the organization that had placed him.

After an interview and a set of selection tests on a remote Scottish island I was told that I would be posted to a school called Chemanza in central rural Zimbabwe, along with three other school-leavers. Just before departing for Africa, I received my A-level results, and I had the grades to study maths at university on my return from my year away. There is absolutely no doubt that my year overseas was formative. It was a remarkable experience, and I will be eternally grateful to Project Trust. It was the first time in my life I had worked hard at anything, and I strove to do a good job. Students that I taught performed at least as well as those in other classes. I grew up a lot in that year, I became much more outgoing and confident, and I hope I inspired some of my students in their educational journey.

During my year of teaching, I fell in love with Zimbabwe and with the people I met. It was also when I decided I didn't want to be a mathematician, and reapplied to university to study biology, which I eventually did at the University of York. As soon as I was back in the UK, I yearned to return to Africa, and the opportunity arose during my second year at university, when I had to choose an undergraduate research project that would be conducted at the end of my second year. It turned out it was possible to design your own project if you could persuade a faculty member to supervise it. I was fortunate enough to persuade a lecturer in the department to supervise me.

A cousin arranged for me to go to Kenya, to the bush camp of George Adamson. My project involved comparing the behaviour of wild lion cubs with three hand-reared cubs whose mother had been shot, and which George was raising for their eventual release to the wild. I spent about a month at the camp, studying lion behaviour.

There were various important experiences during this period that likely contributed to who I am, including the murder of George Adamson and two of his assistants by Somali bandits two days after I left camp. He was the first person I knew who had met a violent death. It was also during this trip that I contracted malaria, which eventually led me to write this book, but I had another formative experience that resulted from this trip that played out on my return to the UK.

Although I had a great time in Africa, I hadn't collected much data for my undergraduate project on lion behaviour. The problem I faced was this. Once I had entered all my data on to the mainframe computer I had access to at university and had conducted analyses, I found that when I drew graphs there did appear to be some differences between the behaviour of the three hand-reared lion cubs and the observations I had collected from wild cubs from a nearby pride. However, when I conducted formal analyses, these differences were not statistically significant. What this means is that the

THE UNIVERSAL HISTORY OF US

differences I observed probably arose by chance. I started to ask myself how many litters of lion cubs I would have needed to collect data from for these results to be significant.

I explored this question by writing some computer code to simulate data. In particular, I used the computer to create a larger data set than the one I had collected, but which had the same statistical properties as the real data. Each column of the made-up data had the same average value as the real data, and the same degree of spread. Some of the columns in the real data were correlated, and the simulated data were correlated in the same way. I would then analyse the made-up data sets and see how big they needed to be to achieve statistical significance. I continued to add rows of data to my simulated data set until an analysis of it reached the threshold of statistical significance.

Much later in my career I learned that there was a much simpler way of solving this problem – an approach that statisticians called power analysis. Although I could have saved a lot of time if I'd known this before, I learned a huge amount about statistics from my simulation exercise. By doing the simulations I taught myself how to code in a computer language called Fortran. I also learned a lot about distributions of data, and about statistical tests. And just as at school I loved learning new things by trial and error, this made me determined to pursue a scientific career. I still get an adrenaline rush when I solve a problem or discover something new. I like the idea that no one else has ever known what I have just found out, however trivial it may be. These breakthroughs are very minor in the grand scheme of things and nothing I have done has had a major impact on the world, but that doesn't lessen the sense of excitement.

Simulating the data taught me more about statistics than

any of the lectures I attended, but it also made me realize that my undergraduate project was deeply flawed. Much in the same way that I cannot prove that a single event radically altered my life trajectory, it is not possible to study three hand-reared lion cubs and a similar number of wild ones and draw any meaningful conclusions. In the same way that no two people are the same, no two lion cubs are equivalent. For my project to have worked I would have had to record the behaviour of many more hand-reared and wild lions, but that was not possible. I described the limitations in the discussion section of my study, submitted my project, and this was enough for me to pass my degree.

Despite being well aware of the limitations of science in being able to assign particular events that happened early in life to personality traits or even the course of one's later life, I still, irrationally, believe the narrative I describe above. I know I shouldn't, but I do think that malaria made me decide to try to understand existence, and that a poorly designed undergraduate research project made me a more competent scientist. Can these narratives really be true? Would I have become a scientist and author if I did not have the mutation in GPR143 and had gone to the Perse school instead of Comberton Village College? Would I have become a mathematician rather than a biologist if I had not gone to Africa? Would I have decided to write this book if I had not caught malaria and nearly died? And did doing a dodgy undergraduate project really set me on a course to become an academic? I can never know because I can't do the experiments required to test these narratives (I'd need many clones of myself), but I do know that all aspects of me are determined by genotype, environment or apparent chance, so perhaps I have constructed this narrative because it includes a little of each.

I have likely decided to believe in the narrative I tell to justify my approach to the science I do or because I wanted a reason for writing a book on what science can tell us about how we got here. As humans we have a strong tendency to look for causes that explain why something happened, and we do this because we want to live in a world we can understand. Most of us don't enjoy change or the unexpected, and constructing narratives helps bring a sense of order to the unpredictable.

The narratives we tell serve two functions. They are used to justify our behaviours and events we have experienced, and we also use them to make sense of the world around us. Our narratives about our life trajectories can contain egregious acts committed by people that have prevented us from achieving our true potential. For example, if Fred Bloggs had behaved more professionally towards my friend Elmer, he would be a better person, would have won multiple Nobel Prizes, and would have wealth and power beyond his wildest dreams. Sure. Sometimes these slights may be real, other times they are imagined. There are very many well-documented historical cases where individuals have acted vindictively to prevent someone else from achieving success or being recognized for it, including in the sciences, and perhaps that is why we sometimes blame others for failures in our lives.

Isaac Newton is arguably the greatest scientist who ever lived. But he wasn't a very nice person. He was paranoid and held grudges, particularly when it came to scientific precedence. He fell out with many other scientists who worked on gravity and calculus, including another great scientist of the era, Robert Hooke. Before Hooke died in 1703, he had been lauded by the Royal Society and a portrait had been painted of him that would help ensure his scientific legacy. Newton became head of the Royal Society in the same year as Hooke's death and he oversaw the relocation of the Royal Society to new premises a few years later, and among the small number of things that went missing were Hooke's portrait and many of his papers. Scientific historians have repeatedly suggested that these losses were not accidental but were overseen by Newton. He was angry, incensed that Hooke had succeeded while alive and accused him of stealing his ideas, and so Newton did what he could to ensure that Hooke was forgotten to history.

Fortunately, not everyone behaves so meanly. Some narratives can contain acts of kindness that are perceived to have altered the trajectory of a life. People who are generally happy with their lot and at home in their skin appear to be more likely to have narratives based on acts of kindness or on their ability to overcome adversity. In contrast, individuals who feel they should be performing better than they are hold narratives where the bad behaviour of others has prevented them from achieving their potential.

I have held the narratives I described earlier for nearly my entire adult life, and they have not changed. But other aspects of the narratives I tell have changed throughout my career. I am currently happy with my lot and am largely content with who I am. The reason for this is that I believe I have achieved more than I expected to in my career. I am in a senior position at one of the best universities in the world, and many of my peers have some respect for me. I am in love with and very happily married to a woman who accepts me as I am and who treats me well. And my children are currently all happy.

During the early parts of my life when I was

unsuccessful in applications for fellowships and university jobs, when I was struggling to make progress on the scientific questions that interested me, or when I was in a marriage that had run its course, aspects of my narrative were different. I did not take as much responsibility for the failings that troubled me as I should have done but instead constructed stories where particular individuals whom I felt had slighted me were out to deliberately hinder my progress. In retrospect, they would have had no reason, nor the power and influence, to do so. My perceived lack of progress towards my career goals was because I was not writing strong enough grant and job applications, and the papers I was submitting for publication were rejected because they were not as convincing to the reviewers as they were to me. Science is a competitive enterprise, and I was not as good at it as I thought I was or as others whom I was competing against. Because some of the narratives I told in the past were false it casts doubt on the ones I still believe and tell. Perhaps catching malaria and simulating data as part of my undergraduate project had no impact on who I am, or my trajectory throughout life.

We know that personal narratives can be wrong. When some people are down on their luck or unhappy with their lot and there is no obvious person known to them that they can blame, they can construct narratives that are easily disproved. These narratives posit the existence of illuminati, shady characters that run global affairs from the shadows that actively hold back and suppress the enlightened who know the truth. One of the more surreal illuminati conspiracy theories posits that the British royal family is descended from humans who mated with a powerful race of malign extraterrestrial reptiles, and this is how they gained their power. A second theory asserts that Prince Harry and Meghan Markle married as part of a devious plan by the British to regain control of America. Specifically, a child of the marriage could become heir to the British throne and President of the United States, providing a route to the British recolonization of America.

Narratives are central to being human and they help us survive in an uncertain world. Science has developed ways of testing many aspects of the stories we tell. The scientific method does this by assigning confidence on whether a cause is linked to a particular event. When the mechanism that links a cause and an event are well understood, confidence can be high. I know that I have a genetic mutation in a particular gene that impairs my vision, and scientists understand that this means that pigment is not produced in cells in the fovea, and that means my eyes wobble. Mutation to this gene repeatedly has the same effect in different people. When science has been unable to determine mechanistic links between cause and effect it does less well and must rely on statistical correlations, and these cannot explain why specific individual events, such as writing this book, occurred. We do not yet have a good mechanistic understanding of how experience influences our personalities or makes us as we are and we must rely on statistical correlations. Experience clearly plays a role in influencing our personalities, but exactly how a particular experience in the past has influenced your, or my, personality is currently impossible to prove.

Because science has some power in deconstructing narratives that are often firmly held, many people are wary of it. Conspiracy theories are easily debunked, yet many people believe them. As an example, a 2019 survey revealed that one in six Britons believe the moon landings were faked, despite having clearly happened. Yet given the ease with which we construct and believe false narratives, it is perhaps unsurprising that scientific knowledge is often disregarded or treated with scepticism. Although science does not have all the answers, it provides narratives that are supported by data and are often testable with experimentation.

Some parts of the scientific narrative I have told in this book are backed up by more data and experimental tests than others, and some scientific disciplines are more mature than others. A mechanistic understanding of the workings of the fundamental forces of nature is quite mature, and this means chemists and physicists rely less on statistical correlations than do those working in biology and psychology, where a mechanistic understanding is frequently lacking. A consequence of this is biologists tend to think of the natural world as unpredictable and rather random, while some physical scientists speculate that the universe may be deterministic and, with enough understanding and computing power, perfectly predictable. Whether we can achieve a mechanistic understanding of the complex chemistry of living beings such as us is an open question, but currently there is a long way to go. Science can't tell me why I chose to write this book, but statistics could tell you that a majority of people want to understand why they exist.

Having told the narrative of the universe and of me, I can now address the question that motivated this book: were you and I inevitable at the birth of the universe, or are we just incredibly lucky? It is not a straightforward question to answer. In researching this book, I have been astonished at the breadth and diversity of human knowledge. I have also realized how flexible the scientific method is, and how researchers in different disciplines use different aspects of it in different ways. I have also learned that although the development of the scientific method is humanity's greatest achievement, it is not perfect. For example, although science can identify that genetic variation, developmental noise and the environment will all have played a role in determining who you are, and that each process plays different roles for different phenotypic traits, science cannot yet tell me why you might be outgoing, empathetic or conscientious. While answering the question that motivated this book, I now also consider how much science has achieved, and how much more there is to learn.

Why We Exist

When I was in hospital in 1989 recovering from my bout of malaria and deciding that by the time I lay on my eventual deathbed my hope was to understand why I had come into existence, I hadn't really thought it through. Scientific knowledge is vast, filling endless libraries, and it grows on a daily basis as more and more papers reporting research are published. Between 2020 and 2023 there were at least four and a quarter million new scientific papers published in English each year. I haven't read them all.

After failing to make any progress towards my life's goal for the first couple of years post-malaria, I finally came up with a plan on how to tackle the Sisyphean task I had bitten off. I started by writing down a list of key things, or entities: energy, force fields, quarks and electrons, neutrons and protons, atoms, molecules, stars and planets, simple life, more complex multicellular life, sentient life, and you and me. I then made a list of the processes that linked each thing to the next on the list. For example, the formation of stars from molecules of hydrogen in nebulae is known as star formation, while the process that resulted in life becoming more complex on our branch of the tree of life is evolution. By focusing on the processes, I realized I could describe what science can tell us of how you and I came to exist in manageable chapters. I could start with how energy became fundamental particles and work through the formation of atoms, stars, molecules, Earth, life, consciousness, humanity and you and me and explain how each thing led to the next. I also came to understand how important interactions were in our existence. I learned how quarks interact to form neutrons and protons, how these interact to form atomic nuclei, and how these, in turn, interact with electrons to form atoms. The atoms then interact to produce molecules, that in turn interact to produce planets, and then – on Earth at least – life. Force fields enabled these interactions. Interactions between competing individuals drove the evolution of complexity, eventually resulting in humans. I realized how the outcome of each interaction was something a little more complicated than the constituent parts. For example, molecules are more complicated than their constituent atoms and protons are more complex than quarks. By the time I had children, I had a plan on how to structure my book. I knew the structure would come at the cost of not describing a lot of scientific knowledge and, in particular, I would skip over a lot of mechanistic understanding.

By focusing on processes, I realized I could address the question of why we exist by describing what had to happen at each stage of our universe's history as one entity on my list transformed into the next. The mechanisms, in contrast to the process, describe how each transition happened rather than why. One reason for skipping over much of the how is that mechanisms can become fiendishly complicated and difficult to write about in a non-technical way. For example, a mechanistic description of developmental biology, a topic I discussed in the last chapter, would require covering how DNA is unzipped, copied and translated into chains of amino acids, how proteins fold, and how different proteins run chemical reactions that play key roles in the construction of our bodies. The subject is fascinating but complex, and a mechanistic description of how you developed from a single cell would run to hundreds of thousands of words, many of which would be technical with very specific meanings. If this book has piqued your interest in any aspects of the science I have summarized, I list some books in the appendix that provide descriptions of the mechanisms that I do not describe.

As I started to plan this book, I delighted in researching and reading about the processes, and the challenge I had set myself became more manageable. I started to gain some understanding of how a simple, minuscule pinprick of extraordinarily hot energy could transform itself over billions of years into you and me, extremely complex organisms built from trillions of cells working together. Yet there were still choices on what to, and what not to, describe. One early challenge was whether to deal with the three laws of thermodynamics. The first states that energy can neither be created nor destroyed but it can be transformed between different forms, the second that systems tend to become more disorganized and less ordered over time, and the third that the level of disorganization eventually settles down to a specific value (I'm not going to explain the value) as the temperature gets close to absolute zero, the coldest temperature that matter can obtain, or −273.15 degrees Celsius. The second law is the most bewildering, given that clearly you and I are highly ordered beings and we have arisen from a point of intense energy over the course of 13.77 billion years. We are not just a random set of molecules that happen to be in the same place at the same time. The history of Earth and of you and me appears to contradict this second law, yet all reputable scientists accept it as correct.

The apparent paradox between the second law of

thermodynamics and our existence is that the law does hold
if we consider the entire universe. On average, the universe
is becoming less ordered, but there are corners of it, such as
Earth, that have become more organized over time. Aver-
ages are useful, but they do not always explain what is
happening in a particular location. While Earth has become
more ordered, with atoms assembling themselves into you
and me, other parts of the universe have become much more
disordered than average. These bits of space will have become
colder, contain very little matter, are much less structured,
and simpler. If we could examine the history of every cubic
inch of the universe, we would find that on average they have
become less ordered, but there is a distribution of change in
how ordered things are, with each bit of the universe having
a different history.

The second law of thermodynamics holds because, on
average across the universe, the degree of disorganization,
or entropy to give it its scientific name, is increasing. I have
written this book focusing on Earth. If I had chosen a point
in deep space, somewhere beyond our solar system but
before another star system starts, the history would be rather
less exciting. I would have described a history of space cooling, the density of energy and matter thinning, and not a lot
happening. The book would have been easier to write,
shorter, but (I hope) not as interesting to read.

I chose not to introduce the laws of thermodynamics
early in the book as I wanted to tell the history of our neck
of the cosmic woods, and I wanted to avoid getting into
discussions of some of the scientific conundrums scientists
have grappled with, or those they still work on. But as I near
the end of the book and turn to the questions of why we
exist, and whether our existence was inevitable or not, I need

to consider some of the hardest questions that scientists, and philosophers, are trying to solve. One of the questions that motivated the book is whether our universe's history was predetermined from its birth, or whether randomness has played a role.

If our universe is deterministic it would develop in exactly the same way each time my repeat universe experiment described at the start of the book was run, as long as the starting conditions were identical. Each universe would end up with identical histories. If the initial starting conditions were to differ between two runs, even by just a tiny amount, the resulting universes would be different. We would consequently be inevitable in a deterministic universe with the same starting conditions as ours. In such a universe, scientists would, hypothetically at least, be able to exactly predict the history of every single particle in our universe from its inception until its demise. Such a feat is technologically impossible with today's computers and is likely beyond the wit of any advanced civilization.

Some scientists argue that the universe is deterministic because they can accurately predict the behaviour of many things, such as the orbits of the planets, and have become better and better at doing so as various scientific breakthroughs have been made. Science progresses and as it does so it increases our understanding. There is only one direction of travel: ever better predictions. For example, 500 years ago we were poor at predicting many celestial events. But today, using equations discovered by Isaac Newton and Albert Einstein, astronomers can very accurately predict the return of comets to our night skies decades into the future, the trajectories of spacecraft, and the orbits of planets around our sun. Similarly, physicists can predict how magnetism will move objects, and chemists can predict what will happen when some types of chemicals are mixed in beakers, in ways that were unimaginable even a century or two ago. Many bits of the natural world are now predictable, and that gives some scientists hope that the future of our universe might be perfectly predictable.

There are many things that we still cannot predict. I can't tell you when my teenage son will get out of bed at the weekend with any accuracy, we can only predict the weather in my hometown of Oxford a day or two ahead with any confidence, and we cannot predict when volcanoes will erupt or earthquakes will shake the ground on which we stand. Given our inability to predict many things, why do some scientists think the universe might be deterministic? Their argument is that we are yet to identify the mathematical equations that would allow us to perfectly predict the currently unpredictable, but that these equations do exist, and it is only a matter of time before we discover them. Scientists have improved our ability to predict all sorts of things, from the way that proteins fold to the existence of new fundamental particles such as the Higgs boson. Scientists will continue to improve predictions, but that does not mean the universe is deterministic.

When scientists are unable to accurately predict things like the weather, they sometimes state that the systems they are failing to predict are stochastic. Those championing determinism would counter that they are not strictly stochastic but that the apparent randomness is due to our failure to identify the equations that accurately describe the system. We might think the weather is stochastic, but this apparent randomness may simply be because we are not very good at predicting the weather more than a day or two into the future. When we discover the right equations, we will be able to accurately predict that 4.2 millimetres of rain will fall on the Champs-Élysées between 9 a.m. and 10 a.m. on 7 January 2081 and we would classify the weather as being deterministic.

One consequence of a deterministic universe, where everything is theoretically perfectly predictable, is that all our actions would be predetermined and we would have no free will. The ways that the atoms from which our brains are built interact with one another would be inevitable, and so too would be each of our actions. Free will would be an illusion in a deterministic universe. I would have been destined to write this book 13.77 billion years ago when the universe began, and I never really made the decision to do so. I could never have resisted the urge to write it, even if I had wanted to. You, too, were destined to buy it.

The idea of free will being an illusion will sit badly with most people, and personally I don't like it. The choices I make feel very real, and I believe we are responsible for our actions. I do not like the idea that our actions might be completely predictable. However, this is just a feeling, and is not scientific. Just because I don't like a hypothesis does not mean it isn't true.

Quantum mechanics very accurately describes the behaviour of the very small, and all the evidence says that aspects of the behaviour of particles such as electrons are random. In my opinion, Occam's razor, the argument that explanations should be as simple as possible yet no simpler, points to a stochastic universe. Once again, such an argument is based on opinion rather than science. These feelings are, of course, important, but they are not scientific. You and I are made from particles, and particles obey the law of physics, so if their behaviour is predictable, then does that mean our behaviour must be deterministic? To disprove the idea of determinism I need to offer an alternative, and the alternative must explain where randomness comes from.

The only true source of randomness so far identified in the universe is described by the theory of quantum mechanics, something I described when discussing the two-slit experiment. Quantum mechanics accurately, yet probabilistically, describes the behaviour of electrons, protons, neutrons and atoms. It is one of the greatest achievements of physics, and explains one of the most peculiar aspects of our universe: that particles are also waves. The theory describes properties of particles as probability distributions. For example, a particle's position is described with a wave function, which describes where it is more likely to be, and where we'd be unlikely to find it. Although we cannot directly observe the wave function of any particle, because measuring it would force the particle to entangle with our measurement device causing the particle's wave function to collapse, we can imagine what it would look like if we could. If we were to scale the quantum properties of particles like electrons to the scale of our solar system, the moon (and sun and all the objects in the night sky) would appear as a distant mist, with some parts of the mist being thicker than others. The thicker parts would be where the moon would be most likely to appear if forced to interact with something.

Quantum mechanics is the most successful theory in science. It very accurately describes how fundamental particles behave, yet despite this accuracy, some physicists argue about the interpretation of the theory. Some claim that the apparently unpredictable and random behaviour of particles will eventually prove to be deterministic, and when that happens free will will be banished to the trash can of discredited ideas. One example of a hypothesis that makes the random behaviour of particles predictable is the Many Worlds hypothesis. Put very simply, and skating over many nuances, this states that we experience one outcome when a wave function collapses, but that all possible outcomes occur in parallel realities that we do not experience. It is a clever way of making the stochastic deterministic by saying all possible things that could happen do, but that we only experience one of the outcomes – our reality.

I didn't mention the Many Worlds hypothesis, or other interpretations of quantum mechanics, earlier in the book when I described the two-slit experiment, because they are currently untestable. There is no evidence for realities beyond our own, and at present there is no way of testing hypotheses that predict them. The Many Worlds hypothesis provides a way to make our universe deterministic and in that respect it is cute. But many creation myths are quite cute too and that doesn't make them right.

Untestable thought experiments such as the Many Worlds theory play an important role in science. They allow us to think through a range of possible interpretations. What I don't understand is why there is a desire among some physicists to prove the universe is deterministic. The available data we have suggests the behaviour of fundamental particles is stochastic, and that opens up the possibility of a stochastic universe and of free will. But it creates a new challenge – how does randomness at the level of the tiny particle translate into non-deterministic behaviour of much larger objects such as you and me?

When the wave function of a particle collapses, such as when a single atom collides with the detector screen in the two-slit experiment, the particle becomes entangled with the detector screen – an object whose wavelength is so small it can be ignored and which behaves like those we are familiar with in everyday life. The screen is designed to detect the particle by interacting with it, and the screen and the particle become a single object. The screen and entangled particle are a large object consisting of millions of atoms, and it behaves like all large inanimate objects in predictable ways. Knock it off the bench and it will fall to the floor, or throw a ball at it and the object and the ball will bounce off it. It is such reassuring, familiar behaviours that make the idea of a deterministic universe attractive to some scientists. But not all solid objects behave quite like detector screens, and perhaps life has found a way to exploit the random behaviours of particles to enable free will.

The field of quantum biology is in its infancy, but exciting insights are beginning to emerge. For example, photosynthesis, the process by which algae, some bacteria and all plants create sugars from sunlight, water and carbon dioxide, relies upon quantum mechanics to transform photons of sunlight into energy that can be used to produce sugars. Photosynthesis evolved early in the history of life, perhaps as long as 3.5 billion years ago, suggesting that life discovered how to exploit quantum behaviour early on.

Evidence is also beginning to emerge that quantum mechanics plays a role in the workings of some enzymes, in the way neurotransmitters work in the brain, and recently biophysicists at the University of Surrey in the UK have data and models that point to some genetic mutations being due to quantum randomness. Genetic mutations influence the development of embryos, and are the fuel for evolution and the spread of life, providing some evidence that we owe our existence to stochastic processes, and that we were not inevitable at the birth of the universe. You and I would not occur in my universe rerun experiment if the universe is stochastic and quantum processes influence evolution.

Some claims of quantum biology are still controversial, and more evidence will be required to convince sceptical scientists. Many biological molecules are too large to behave like single atoms, and some researchers have argued that the warm, wet world inside our cells would prevent life from exploiting the quantum properties of the very small. Nonetheless, given our biological understanding that interactions between pairs of molecules impact many cellular processes, and that this is the scale at which random quantum interactions happen, there is potentially a route via which randomness could make our existence down to chance. A number of scientists now argue that the warm, wet environment inside cells is ideal for quantum processes to scale up to influence the dynamics of whole organisms. However, this does not answer how quantum mechanics might result in consciousness and free will. At this point, the evidence becomes sparse and I am straying into conjecture. The Many World proponents may now see an opportunity to accuse me of generating an untestable hypothesis. But I would argue that my theory is more testable than theirs.

In the chapter on the evolution of consciousness I described how the neocortex processes information. The neocortex, a highly folded part of the brain, is very large in humans, and is the part of the brain that makes us intelligent. Inputs to the neocortex from our sense organs – the eyes, ears, nose, skin and tongue – lead to a propagation of signals through networks of cells until neurons fire that allow us to recognize particular objects: remember the Jennifer Aniston neuron. Neurons in different clusters across the neocortex may fire, with signals from multiple neurons processed before a consensus is reached about what we are experiencing. Other neurons then propose how we should respond and a decision is made.

Although data are largely lacking, a number of psychologists, mathematicians, philosophers and physicists have proposed a role for quantum mechanics in consciousness, and given that sentience appears to arise from electrical activity of neurons interacting via synapses, there is certainly potential for stochastic quantum processes at the level of interacting molecules to contribute to consciousness, and ultimately to free will. Exactly how is currently a matter of speculation. Nonetheless, we cannot rule out a role for quantum randomness across the hundreds of thousands of synapses that fire every second that contribute to the way we make decisions. I have strayed from what science can currently tell us into guesswork, so let me bring it back to a line of evidence that is a little more concrete.

Many scientists and philosophers have drawn comparisons between computers and the human brain. They usually point out how aspects of the way computers operate mimic the workings of bits of the human brain. Such comparisons are made more and more frequently as artificial intelligence becomes more embedded in our everyday lives. Many of the algorithms that computers use to make choices when playing a game such as chess, or to find optimal, or near-optimal, solutions to currently analytically intractable problems, rely on random numbers. For example, ChatGPT relies on random numbers when you ask it to complete a task and computer scientists use randomness to help solve all sorts of computational problems that, today at least, cannot be solved without recourse to chance. Randomness can be used to identify solutions to problems, and if genetic mutations are due to chance, as seems probable, then we owe our existence to random events. It might seem counterintuitive that randomness can help evolution find solutions to make organisms more competitive, or computer scientists can use randomness to find optimal solutions to problems, so I will briefly give an example of how I have used random numbers in my research to solve problems.

My Ph.D. research investigated the impact of seed predators and herbivores on tree reproduction in forests in the US and the UK. In hindsight, it didn't move the field forward much, but I don't think it set it back at all either. I did learn a lot, and one thing I learned was how I could use a random number generator on the computer to construct a statistical model of seed and seedling predation by squirrels. I had collected quite a lot of data from experiments where I placed acorns in various densities across the forest floor, some in deer- or chipmunk- or squirrel-proof enclosures, others not. I would then return to see how many had been eaten. I then wrote down a mathematical model to describe the impact of vertebrate seed predators on the likelihood of seeds in different places germinating. The next job was to combine the mathematical model and the experimental data to identify the values of parameters that gave me the model that did best in explaining variation in the data I had collected.

I wrote a program in a computer language called Turbo Pascal to find the most likely values of each parameter in the model. The program implemented something called the Metropolis–Hastings algorithm that uses random numbers to hone in on the most likely solution. The way the algorithm works is akin to finding the highest point in a mountain range in thick fog at night using a teleportation device and random numbers.

At each location in the finding-the-highest-point analogy you can measure the altitude. Having done so, you randomly select a direction to move in and then teleport a fixed, pre-determined distance before measuring the altitude at this new location. You then compare the old and the new altitude. If the new location is higher, you move to it 70 per cent of the time, but if it is lower, you stay put in your original position, also 70 per cent of the time. You do this again, and again, exploring the mountain range. In the Metropolis–Hastings algorithm you are exploring a likelihood surface. Instead of latitude and longitude coordinates of the surface of our planet, and an altitude, the lats and longs are replaced by parameter names – perhaps alpha and beta – and altitude is replaced with a statistical quantity called the likelihood. The likelihood describes how well the model predicts the data you have collected given the values of alpha and beta. The highest value of the likelihood is called the maximum likelihood and it is analogous to the peak of the highest mountain in a range. The Metropolis–Hastings algorithm is one way to explore the likelihood surface and find the maximum likelihood.

The mountain-range analogy breaks down a little when a model has more than two parameters. If a model has, say, four parameters called alpha, beta, gamma and delta, instead of the latitude and the longitude coordinates on our mountain range analogy we now have four dimensions to explore instead of two – one for each parameter. With altitude, the latitude and longitude coordinate produce a total of three dimensions, while with four parameters and the likelihood my example model has five. I cannot visualize a surface that has more than three dimensions, but mathematically it does exist. I find it much easier to imagine a mountain range, and that is why I use the analogy. The Metropolis–Hastings algorithm works in any number of dimensions, allowing scientists to find the values of all the parameters in a model that maximize its likelihood.

There are two tricks in getting the Metropolis–Hastings algorithm to work. First, you need to select the distance to move such that you can explore the entire likelihood surface. If the distance between where you are now and the next place to sample is small, all that happens is you end up on the top of the mountain you start out on. You might be high, but there may be taller peaks in the range, which means you won't be at the highest point. In maths-speak, you would be at a local maximum rather than the global one you are looking for. Second, how often should you not move to a higher location? The 70 per cent value works well, but other acceptance rates may on some occasions be slightly better (it depends on the specifics of the problem being addressed).

Assuming the distance to move and the acceptance rate are appropriate, the Metropolis–Hastings algorithm is a faster way of finding the best solution to a problem than deterministically evaluating the likelihood at each point on a grid laid across the likelihood surface and mapping the entire mountain range. The algorithm is also only one of very many that computers use to solve all sorts of problems in statistics and artificial intelligence. Computers are often described as mimicking the human brain, and perhaps we too use randomness to make decisions on what to do and how to behave. The only known source of randomness we could use is quantum randomness at the level of the fundamental particles, atoms and molecules, but perhaps science will one day uncover other sources of randomness in the universe.

Evolutionary biologists use a similar analogy when

thinking about how evolution can produce more competitive individuals, lineages and species. Instead of being parameters, latitude and longitude are now values of phenotypic traits, and altitude is replaced with how good individuals with particular trait values are at surviving and reproducing. Random mutation means some individuals are less good at surviving and reproducing than their parents because they have less fit phenotypic trait values. However, every now and then a mutation generates phenotypic trait values that increase survival and reproductive rates. Evolution slowly climbs towards the survival and reproductive peak, but in doing so produces individuals that struggle to survive or are unable to reproduce.

Randomness is potentially pervasive through biology, yet research into its role is still in its infancy. Quantum biology and quantum consciousness are very new fields of scientific endeavour. If I were starting out in science today, I think I would seriously consider immersing myself in one of these fields. I reached this conclusion, because in researching this book, I have formed the opinion that until physicists are able to produce data to demonstrate that quantum mechanics is attributable to some underlying deterministic process, then I will take the quantum strangeness at face value, and conclude that the universe is stochastic. If I was to rerun my replicate universe experiment, each outcome would be different. Perhaps I would find the physical universe is deterministic, and our solar system will exist in each universe rerun, while life is stochastic. Maybe sometimes when we visit Earth in each universe, humans will exist, while at other times perhaps a species of malevolent intelligent lizard is the dominant species on our planet. Or perhaps I am just trying to find a way to bring the physicists who argue for a deterministic universe and the biologists who are convinced the universe is random together. Regardless, this discussion does not answer the question of why we exist.

I started my personal journey to try to understand why I existed when I lay on my deathbed. Writing this book does not, I hope, signal my imminent demise, as there is still much more I want to understand. Nonetheless, I have some idea of why it is we exist.

One downside of the universe being stochastic, and the Earth being the only planet where we have been able to study life, is that it is impossible to confidently assign probabilities to events that we only have evidence of happening once here on Earth. In particular, Earth is the only planet on which life is known to exist, so assigning an accurate probability of life evolving on other planets, even if they are similar to Earth, is currently impossible. We need more data before we can state if life arises frequently or not.

Despite this challenge, scientists have made remarkable progress in working out what had to happen for us to exist, why it happened, and even how it happened, but there are a few key events in our history we can still only speculate about.

The first key event we cannot assign a probability to is the birth of our universe. We have never observed another universe, and we have never observed anything beyond our universe. We don't even know where the edge of our universe might be, or indeed if it even has one. We cannot look beyond our universe in an attempt to observe others coming into existence (or not). We consequently do not know whether universes are common or if there is only one, and if they are common, whether ours is typical or unusual. Put another way, we do not know why there is something rather than nothing.

What we do know with reasonable confidence is that when our universe formed, it consisted of a singularity of extremely hot energy. The four fundamental forces quickly emerged as the singularity expanded and cooled, and the fundamental particles from which all matter is made came into existence. Much like the formation of the universe, the four forces emerged only once, so we do not know whether it was inevitable or fortuitous that gravity, electromagnetism and the weak and strong nuclear forces emerged with the strengths that they have. Perhaps a different set of forces would emerge if we were to conduct the rerun experiment, and there may be fewer, or more, than the four we observe in our universe. Another unknown is consequently why we have the forces we do.

What we do know is that life as we know it requires all four forces, and computer simulations reveal that even small differences in each of their strengths could result in universes without atoms, molecules, stars or planets. Ultimately the reason we exist is that the fundamental forces took the values they did, and this allowed protons, neutrons, atoms, stars, molecules, galaxies and planets to emerge. We are here because the fundamental forces are just right. Given they have these values, science has a very good, but not yet complete, understanding of why these things formed and why they behave as they do.

Some scientists and philosophers have marvelled that our universe has forces of the right strength for life to exist. Some even interpret this coincidence as evidence that our universe could not have arisen by chance. However, if our universe did not have forces that were just right for life, then life could not evolve and no one could observe it. We can only observe the universe because it is just right for life to evolve. There may be many other universes that weren't right for life, and they have never been observed. The fact we can observe our universe because things were just right for life to evolve is called the anthropic principle. We exist because the early universe developed in a way that would become favourable for life (at least in one location). We don't know why this happened, but we should not be surprised, because if our universe hadn't developed like this we could not exist or wonder about our existence. I think there are more exciting questions to ponder than the anthropic principle, which I find a circular argument.

Unifying the standard model of electromagnetism and the strong and weak nuclear forces with Einstein's theory of the way gravity works would provide a more complete under-standing of our universe, and is an exciting research area in theoretical physics. Theories such as string theory and quantum loop gravity provide approaches to unify the four forces, but testing them, and showing which theory is right, requires so much energy it is currently beyond our technical exper-tise. Nonetheless, Einstein's theory of general relativity and the standard model provide good explanations for why matter behaves as it does. In principle, large particle accelera-tors might one day be able to re-create conditions a little more similar to those found in the early universe than we can achieve today, and experiments in these machines might shed light on why and how forces emerge with particular proper-ties, but currently our technology is not sufficiently advanced to build such a facility.

If I was to create universes, each with our gravity, electro-magnetism and strong and weak nuclear forces, where quantum behaviour is truly random, the universes would look similar, but not identical, to ours. In any universe with identical forces to ours, physical and chemical processes will be the same as those uncovered by scientists in our reality. The first generation of stars would consist of hydrogen and helium, and their deaths would create heavier elements in the ratios we observe. The molecules we are familiar with would also form. Water, carbon dioxide, nitrogen, methane, gold, uranium and other elements and molecules would be as common as they are in our universe. The fundamental forces determine the dynamics of all elements and molecules, so as long as these forces emerge as they are in our universe, and quantum behaviour is random, the processes we observe in our universe will be identical to those in our experimental reruns, yet each universe would be unique. We could exist, but we wouldn't necessarily do so.

It would be likely that our solar system would not exist in other universes in my rerun experiment if our universe is stochastic. The early universe was not completely uniform. There were small variations in the intensity of energy, and in the density of the fundamental particles that formed. Some parts of the very young universe consequently had more quarks and electrons than others, and this variation translated into the distribution of galaxies across the night sky. Mathematical models suggest this early variation in the distribution of matter might have been caused by quantum randomness. In a stochastic universe, such randomness would result in a different layout of galaxies each time a universe like ours formed. There would be the same distribution of small and large galaxies, spiral and elliptical ones, but each galaxy in each universe would be different. Only our universe would have the Milky Way, the sun, the Earth and the other planets in our solar system.

Although unique, the Milky Way is not unusual. It is a

galaxy of average size. Similarly, our solar system is one of a kind, but it too is not odd. Our sun is only a little larger than the average star, and observations of other stars through telescopes reveal that the planets that orbit our star are not particularly unusual. NASA reports that at least one in six stars has an Earth-sized planet, but some of these are orbiting stars that are very different from our sun. Another study has revealed that about one in five sun-sized stars have Earth-sized planets in their habitable zones. In contrast, only about one in twenty stars that are the size of our sun are orbited by gas giants like Jupiter and Saturn, planets that played such a key role in determining Earth's current orbit. We can conclude that systems like ours, with planets like Earth that might be able to support life, are not particularly unusual in our galaxy, and we would expect them to occur at the same frequency in other galaxies in our universe.

Assuming four forces of the right strength and a large universe, Earth-like planets orbiting the right distance from sun-like stars for life to form appear to be quite likely. However, the suitability of Earth for life is dependent not just on its position in the habitable zone but also on its orbit and speed of spin. We currently do not have the data to study these attributes of other Earth-like planets, although scientists have proposed ways of doing these studies, so the data may be available in the not-too-distant future. But as I write we do not know whether our twenty-four-hour days are typical or unusual. Mars, our closest planet, has a day length very similar to that of Earth, but the next closest rocky planet, Venus, has a day length of 5,832 hours. Much more work is required, but the number of Earth-like planets suggests that some must have orbits and rates of spin appropriate for life.

The moon is also important for life on Earth, helping

stabilize the climate by reducing our planet's orbital wobble. We do not have the technological refinement to count moons around Earth-like planets in other star systems, but we do know they are common around the planets in our solar system, with over 200 moons counted. They can form at the same time as the planet, be captured by planets having started their lives as meteors and comets or even moons of other planets that escaped their orbits, or via a collision, as happened to produce our moon. We are the closest planet to the sun to have a moon, and our moon is the fifth largest in the solar system. Our moon may yet prove to be a little unusual, but it seems likely that some Earth-sized planets around other stars could well have one or more moons.

Part of the reason we exist is that the Earth has the right orbit, the right day length, the right axial tilt and a moon. Are these properties unusual? More observations of the heavens will help answer this question. Even if we conclude that close twins of Earth are rare, the cosmos is so large there may still be hundreds of thousands, or millions, of them in our galaxy alone. Yet life may not need an Earth-like planet to get going. Life could potentially arise on planets very different from Earth, or even on moons that harbour liquid water. Enceladus, a moon of Saturn, and Europa, a moon of Jupiter, both have salty oceans of liquid water below a thick layer of ice. Mars had surface oceans similar to those found on Earth today, although by 3 billion years ago they had been lost to space. Another four or five moons of the outer planets potentially have oceans trapped below a thick layer of ice. Could life have evolved on an ancient Mars or on these far-flung moons? We have no evidence that it did, but we have also been unable to look in any detail, so what can we say about life? Is the universe teeming with life, or is restricted to Earth?

The building blocks of life – nucleobases, amino acids and other organic compounds – readily form in space as revealed through analyses of meteors, meteorites and space dust. These molecules also form on Earth. The key molecular building blocks needed to get life started are consequently common, but that does not mean life routinely springs into existence – an energy input is needed. Volcanic energy sources likely powered the emergence of life on Earth. Volcanoes are not unique to our planet. Volcanoes, some dead, have been observed on Mercury, Venus, the Moon, Mars, and Jupiter's moon, Io. Quite how volcanic energy and organic compounds combine to produce life is unclear, although scientists have plausible hypotheses of how life got started. However, they have not been able to identify the requisite conditions for it to get going in the lab. Nonetheless, membranes readily form, autocatalytic chemical reactions are not uncommon, and simple metabolism seems inevitable. What we don't know is how easy it is for these key components of life to combine.

Although we do not know the conditions required for life to get started, there is no reason to suspect that the conditions on early Earth were particularly unusual. We have no reason to suspect that the emergence of life is a rare phenomenon. Of course, we cannot rule out that something extraordinarily unusual is required for life to get started and that only happened on Earth, but in the absence of any evidence of this, it seems reasonable to conclude that primitive life is a common feature of the universe. Whether it always uses RNA and DNA is unknown. Perhaps self-assembly manuals can be written in other molecular codes, but given amino acids are found on comets and meteors, my suspicion is that life will usually use RNA and DNA for coding instructions on how to build a cell. But much as French, English and Spanish use very similar alphabets to form a different language, the same base-pair triplets will not necessarily code for the same amino acids in independently evolved DNA-based life. Cytosine–adenine–guanine codes for glutamine on Earth, but it might code for valine or serine elsewhere. Or perhaps a completely different set of amino acids will be used.

Once life begins, competition between individuals, either of the same or of different species, is inevitable. The competition will always be for resources, and it leads to the evolution of life forms that can thrive on a wider range of resource types or exploit new resources. Evolution will always select for individuals that are more efficient at detecting, acquiring and utilizing these resources. Competition between life forms (where they exist) will be pervasive throughout the universe as the fittest replicator will always displace those that are less fit. On Earth, this competition has led to an increase in diversity and complexity in the eukaryote branch of the tree of life. As life diversifies, some of it may inevitably increase in complexity, as competition from a more diverse array of competitors will select for strategies to thrive in ever-more complex environments. Once life gets started, I suspect it will routinely proliferate and increase in complexity.

The next question on the journey from the Big Bang to you and me concerns whether multicellular life will inevitably evolve. First, multicellularity on Earth requires an oxygenated atmosphere. Oxygen will be a waste product of most forms of early life that use chemosynthesis and photosynthesis to power metabolism. It is also extremely reactive, and free oxygen cannot accumulate in the atmosphere until elements or molecules it easily reacts with have been oxidized. Planets are large, and the oxidization of surface molecules takes time. On Earth it took about 1.5 billion years before atmospheric oxygen began to increase.

Multicellular life required atmospheric oxygen to evolve, but once oxygen levels started to rise it evolved at least twenty-five times. Eukaryotes needed mitochondria to use the oxygen to run their metabolism. Mitochondria evolved from bacteria that were able to survive inside other cells. Predation or parasitism had to evolve to allow these bacteria to enter other cells before they were co-opted into a mutualism by the host. Over time, these bacteria evolved into organelles such as mitochondria, that are used by all multicellular organisms to run their metabolism. Evolution is very good at producing organisms that can exploit abundant, under-utilized resources, so it is possible that predation and parasitism would evolve given enough time on other planets. It also seems probable that once predation and parasitism evolved, mutualisms with bacteria living inside host cells could also arise. All eukaryote cells contain mitochondria, derived from a mutualism with aerobic bacteria, while plant cells also contain chloroplasts that evolved from photosynthetic bacteria such as the cyanobacteria. Given such mutualisms evolved independently at least twice on Earth, it seems plausible that they could evolve elsewhere too. My hunch is that multicellular life will almost inevitably evolve from simpler life forms given sufficient time.

On Earth, a lead time of over 2 billion years was required before multicellular life evolved. Our planet had to remain habitable while avoiding cataclysmic extinction events such as local supernovae, meteor impacts, or the loss of atmosphere from solar winds that could wipe out all life. Earth did this, but we currently have no idea whether this is typical or not. It seems likely that once life evolves, it inevitably becomes more complicated, but life is fragile, and is permanently at risk of being killed off by violent cosmic events. We exist because Earth remained habitable for 4 billion years, allowing evolution to work wonders, but we do not know whether such durations within habitable zones are commonplace or rare.

Once complex multicellular life had evolved on Earth, there was no guarantee that humans would appear a little over half a billion years later. Random, unpredictable events driven by mutations mean that evolution is unpredictable. In an experiment with the bacterium *E. coli*, Rich Lenski allowed twelve colonies of bacteria to adapt to the same environmental conditions. At the time of writing, the experiment had been running for 75,000 generations. Initially it looked as if each population was evolving in a phenotypically similar manner, but examination of the genomes of the bacteria revealed that different colonies accrued different mutations. In other words, although phenotypic adaptations initially looked similar, they were not achieved via the same genetic mutations. Such a pattern is not unusual – there are often multiple genetic routes to the same phenotypic solution.

The most remarkable adaptation occurred in one of the twelve colonies shortly after generation 31,000. In the presence of oxygen, *E. coli* use glucose as a primary source of energy to run their metabolism, but they can also run their metabolism off a compound called citrate in the absence of oxygen. *E. coli* in the laboratory are typically fed a diet of glucose and citrate, but will not utilize the citrate in the presence of oxygen, and this was the diet that was fed during Lenski's long-term evolution experiment. One colony evolved to utilize citrate in the presence of oxygen. It is difficult for *E. coli* to do this because it involves a series of mutations and not just a single one. These mutations occurred only in one colony. The bacteria in this colony can now use a resource unavailable to the other *E. coli* colonies. Over longer periods, phenotypic evolution is not repeatable, and there is no guarantee a particular life form will evolve.

Such rare mutational events, perhaps driven by quantum fluctuations, will have played occasional but important roles in the evolution of life on Earth. It is highly improbable that life on other planets would experience the same set and order of random mutation events as occurred on Earth, particularly if it uses different triplets to code for amino acids. Scientists do not know enough about the genetic architecture of most phenotypic traits, or about which phenotypic traits are most likely to arise, and which are less likely to evolve. Life on Earth today is a combination of what has gone before in terms of mutation and selection. Although natural selection can frequently be repeatable, genetic mutation is less repeatable, and this can lead to quite different outcomes over millions of years. Deer and kangaroos have similar ecological roles respectively in Europe and Australia, and they are both capable of thriving on vegetation, but they look very different. The evolution of multicellular life on other planets may result in some outcomes that look familiar to us, but others that might look very odd. Aliens really could be little green men.

Once multicellular life arose, it seems inevitable that evolution would find ways to detect light, sound waves, aromatic molecules and touch, and this would require a central nervous system and some form of brain. Consciousness would then naturally emerge, and in time intelligence would appear. Intelligent life arose on Earth in the form of humans, and their close cousins, the Neanderthals. Other, less closely related species, including dolphins, elephants, crows and octopuses, stand out as also being able to solve quite complex tasks. Although complex language and technological development have evolved only in primates, this does not preclude it from evolving in other species.

Several writers and scientists have concluded that life, and intelligent life in particular, are likely to be rare. I don't believe we can make that call yet, as we just do not have enough evidence. Most things in our universe, be they galaxies, stars or star systems, have distributions. For example, galaxies may contain a thousand stars or as many as 100 trillion. Their size varies enormously, with the Milky Way lying somewhere in the middle of the distribution. In the absence of evidence, the most sensible thing to conclude is that what you are observing is more likely to be average than not. Our galaxy is average, the sun is average, and rocky planets in the habitable zone are also not at all unusual. It may turn out that Earth is quite typical and that life is abundant. We will not find it on every planet, or even around every star, but if I were forced to guess, I would say that life evolves frequently on rocky planets with liquid water and volcanic activity, and that we are far from the only planet where intelligent life has emerged. Life is chemistry, and chemistry follows sets of rules, so as long as conditions are right, life will emerge.

As a species, humans are very probably unique in the universe. I cannot see how species with the same genetic code as ours evolved to live on planets around other suns. Genetic mutations occur at random. We are a product of genetic mutations raining down on our ancestors' genomes for 4 billion years. An identical sequence of random events will not have happened on another planet, anywhere, so as a species we will be unique.

The development of technologically advanced civilizations will require easy access to energy. Without it, we could not make accessible the large amounts of information necessary to develop and manufacture telescopes, computers and particle accelerators. We built our civilization on coal and, later, oil, and I doubt we could have developed it on tide, volcanic or wind power alone. How common will easy access to energy such as coal be? It is possible that advanced alien civilizations built on fossil fuels are rare, although once again we lack evidence to state this with any confidence. Coal formed on Earth because bacteria and fungi were unable to access the dead plant material from which it formed as a resource, and it consequently formed the hydrocarbons we have built our civilization on. Evolution is a potent force, but it requires genetic mutations that appear at random. Coal formed because the genetic mutations did not happen that would have allowed organisms to use the dead trees of the warm Carboniferous as a resource. If the development of advanced civilizations is dependent on chance genetic mutations, they may be rare. We know how evolution works and why it produces new life forms, and we know we are an accident of evolution. So too may be our civilization, and this means that perhaps very few species achieve an understanding of the universe that is as well developed as ours.

My personality, like yours, is also an accident of history, even if I cannot tell you with confidence which historical events made me the way I am. I have experienced a unique set of events and experiences, and these have helped make me. My personality, my desires and my obsessiveness to understand why I exist are due to a mix of my genes and the environment I have experienced throughout life. Had I been born into Tudor England, Napoleonic France or even ancient

Athens, I would likely have developed other desires and motivations. I believe my brush with malaria was key to me being who I am, but I cannot prove that. Despite being an accident of history, scientists have some idea why I have the personality I do: chance, nature and nurture will all have played a role.

I exist because the fundamental forces have the right values, because these forces produce galaxies, stars and planets, and because chemistry is such that life can evolve when conditions are right, and evolution can act to make it complicated and intelligent. I am persuaded that our universe is stochastic, and this stochasticity manifests at the level of fundamental particles, and complex living organisms have free will because life has found a way to exploit this quantum randomness at the level of living beings. Despite this there is still much I want to know before I lie on my eventual deathbed. I want to know why there is something rather than nothing, and what the conditions were that led to life beginning. I doubt scientists will solve the first question in my lifetime, but I hope we will be able to make simple life in the laboratory by the time I die. It may be a long shot, but we are making progress. I know many of the reasons why we exist, but not all. Science has brought us a very long way, but the history of why we exist is not yet complete.

I can't write a book about existence without touching on unscientific explanations for us being here. Many people believe in deities, life forces or cosmic energies for delivering their existence. Given the prevalence of these beliefs, can they all be wrong? I think so, but that doesn't mean I don't respect people who hold other beliefs. Perhaps that is just part of my personality and can be attributed to some childhood event I have long forgotten.

Many non-scientific beliefs can easily be shown to be non-sense. An acquaintance of mine is convinced that crystals have some supernatural properties that protect from malevolent spirits out to cause harm. They can protect themselves from such dangers by always having a crystal in their underwear. Such patent nonsense doesn't bother me enough to attempt to disprove it, but the experiment that would do so is easy to design. I simply need a large group of volunteers that I would randomly assign to two groups. I would ask one group to wear underpants with crystals sewn into them for a year, and the other one to wear pants that are identical in every way but lack the crystal. I would score each participant in the experiment for mental and physical well-being at the beginning and the end of the study and calculate changes in these scores over the course of the experiment. If crystals provide some health advantage, then those in the crystal-lined-underwear category would on average improve their health to a greater extent than those with less ornate pants. I am confident there would be no difference, with the possible exception of increased chafing in the crystal-wearing group. Related experiments have been conducted, and crystals provide no health advantage.

Other equally wacky ideas are yet to be disproven, but there is also no evidence to support them. There is no evidence of any force beyond the four fundamental forces permeating the universe. There is no life force or positive energy field. There is also no scientific evidence for any of the deities that humans have created to explain their existence, but that is part of their attraction for many. Deity beliefs are impossible to disprove because they are constructed around blind faith. Believers are rewarded, typically after death, for holding evidence-free beliefs and behaving in a particular way while alive. Such narratives are constructed in a way that they cannot be disproved by science, and this is why science and religion often clash. Many scientists are critical of religion because science is evidence-based while faith is not, while the devout often see science as a threat as more people turn to it for an explanation for their existence. Despite this, religion is appealing to many because it gives life purpose in a way that science does not, and because we all feel special.

At about the age of twelve or thirteen I started to feel like I was special. I thought that I was destined to do something great in later life, perhaps become prime minister or Britain's greatest-ever Olympian. These feelings didn't motivate me to become immersed in politics or sports training, it was just a feeling of inevitability I had. It felt that my soul had been assigned a body, that this gave me an opportunity not afforded to all souls, and I had some sort of duty to make my life worthwhile. At this time, I believed in the Christian God, having been educated at a Church of England school and being sent to Sunday school as a child. My mother has been a regular churchgoer all her life, and religion is important for her. To her great credit, she has never criticized me for my lack of faith, and we have always respected one another's views. Along with many others she will likely disagree with the following paragraphs, but she will be proud of me for writing this book.

By my mid-teens, as I became ever more fixated with science, I began to lose my religion, and by the time I had finished my undergraduate degree I was an atheist. Part of the challenge of that journey was, if my life was to have a purpose, what was it going to be? My brush with malaria helped me decide.

My experience of feeling special is not unique, and I suspect is an inevitable consequence of consciousness. When pondering the question of why they exist, many people interpret this feeling of being special as assuming they have a higher purpose. They accept that they exist, they feel (and they are) special, and so there must be a reason as to why they are here. These readers may feel short-changed in that I have described what had to happen in the last 13.77 billion years for them to exist, for I have not given their existence a purpose. Apart from the evolutionary argument that the purpose of their existence is to attempt to reproduce, I argue there is no other reason. Our existence is a consequence of the strengths of the fundamental physical forces, evolution and a lot of randomness.

I also don't understand why, if our existence does have a purpose beyond genetic replication, it took nearly 13.77 billion years for us to arise. What were deities up to for all this time? And why are we located in a rather pedestrian location in an average galaxy rather than at the universe's centre? Surely a deity would have chosen a more glamorous bit of cosmic real estate? Given our tiny place in the vastness of the cosmos, it seems arrogant to suggest that life on Earth has been attributed some higher purpose. Are we as a species unique in being assigned a purpose? Or was each Neanderthal endowed with a purpose? Evidence suggests they had religion. Perhaps it was one of our earlier ape ancestors that God gave purpose? Or what about the first species to emerge from the sea to forge a life on dry land? Or the first bacterial cell? Perhaps all the living are special and given a purpose by a god, but that does not separate us from the rest of life.

The lack of a purpose-focused answer to the why-we-exist question does not mean we should live lives without meaning. I have chosen to spend my life trying to understand how I came to be, but this is a personal choice rather than one ordained. I have no doubt some reviewers will think I have wasted my time, but that doesn't bother me as it was a choice I made. My life is not lessened by having no preordained purpose. On the contrary, I find the wonders of the universe significantly more inspiring than the story told in the Book of Genesis – the first book to attempt to tell the story of our existence. The fact that I am lucky to be alive does not diminish my enjoyment of my days on Earth.

One of the key reasons I abandoned my faith was the teaching that the Christian God was omnipotent and good. The existence of suffering, inequality and people such as Hitler and Lenin was impossible for me to reconcile with the God I had been taught about. To me, a better explanation for the horrors of the world is that they are not preordained. I developed a form of atheism that puts my existence down to chance and good fortune but which can also explain why suffering and bad behaviour exist.

If we are just another species, arising as a consequence of chance mutation and selection acting over billions of years, I wondered what it meant for humanity, morality and ethics. Our existence, if it is down to luck, does not mean humans, or you and me, are uninteresting and unimportant. On the contrary, humans are an unusual species, being the only one (on Earth at least) to have built cities, mastered fire, written novels and conducted scientific experiments. Our brains make us conscious, and this, coupled with our abilities to communicate complex ideas, to envisage possible futures and to trade off current actions with potential immediate and future consequences, sets us apart from the ants, anteaters and antlions. The ability to envision possible futures is important, in part because it brings with it the realization that our deaths are inevitable, and this in turn raises the question of why we exist.

The biological explanation of why we exist – to reproduce – at first glance lacks humanity. I exist for the same biological reason that all *E. coli*, earthworms and rose bushes do. My ancestors were successful at reproducing, and by definition my descendants require me to be good at it too. Where do ethics and morality come from given the reductionist explanation for existence I champion?

Part of the answer could be in who we choose to mate with, an idea initially proposed by Charles Darwin, the discoverer of evolution. If people tend to choose partners on aspects of their character that are determined, at least in part, by genes, as long as these more favourable partners have more children than everyone else, their desirable characteristics will become more prevalent in the population in future generations. Although far from conclusive, there is evidence that moral views can be heritable, so the fact that you are scrupulously honest while our past prime minister who delivered the Brexit agreement was less so may in part be due to genes.

There is no compelling evidence that people who are more honest, moral or ethical have more children. However, widespread access to birth control in countries where most of these studies have been conducted makes the study of the evolution of morality in modern societies challenging because the number of offspring a couple produces has become a matter of personal choice rather than genetic make-up. Nonetheless, perhaps our ancestors did have behavioural characteristics that differed from their peers who have no descendants living today.

There is another route via which morality and particular

types of behaviour can evolve. We are a social species, living in groups. In most group-living species, it is important to be a group member, with solitary individuals often at risk of an early death, or of a failure to find a mate. A solitary meerkat, for example, is unlikely to survive for long. Throughout much of our evolutionary history, failure to be part of a group would have been risky, yet because we can communicate complicated ideas with one another and consider the future, groups of our ancestors could make decisions about membership. Individuals could be banished from groups, or even killed, if their actions were deemed unacceptable by the rest of the group, or even a powerful minority of it. If badly behaved individuals were more likely to be evicted from groups by our ancestors and meet an early death than those who exhibited less selfish behaviour, and if these bad behaviours were in part due to genetics, these behaviourally deleterious alleles would have become less common over time. In the last few years I have worked to help the community of biologists in my department in Oxford. Doing this has taken me away from research, something I enjoy. I did not make the decision to try to help researchers in my department because of a belief that I should do good. I did it because I felt I was in a position to help. I did enjoy building a stronger community of biologists. And I didn't need to believe in a deity to sacrifice my research to try to help others.

We cannot solely blame genes for unethical or morally reprehensible behaviour. Like most of the phenotypic traits I have discussed in this book, environment can also play a role in their development. Some groups may be more accepting of certain types of behaviour than others, because the cost of certain behaviours to a group may differ depending upon where it lives. Stealing trinkets made from readily available seashells may have been tolerated by beach-dwelling groups, but if similar trinkets were stolen within groups that lived inland where shells were rare, these thefts could have been considered intolerable. I have chosen a relatively mild form of antisocial behaviour to make this point rather than focusing on more violent acts, yet similar arguments can be made for coercive and controlling behaviours. I did not wish to detract from the point I wished to make by focusing on a more shocking behaviour.

Evolution occasionally throws up abnormalities. I described earlier in the book how I have a genetic mutation that negatively impacts my eyesight. Extreme instances of any phenotypic traits determined by genes can arise through mutation, and this includes some forms of antisocial behaviour. Every now and again, individuals that do not adhere to standard cultural norms develop, with both genetic and environmental factors playing a role. For example, mutations in a gene called MAOA appear to be associated with psychopathic behaviour. I have no idea whether Charles Manson, Jack the Ripper and Vlad the Impaler were genetically predisposed to psychopathic behaviour, and even if they were this does not excuse the murders they committed, but it is possible that genetic mutations played a role in their dreadful crimes. There is speculation that Hitler, Stalin, Putin and other murderous dictators may also have had mental disorders.

In evolutionary explanations for humanity, ethics and morality, it is inevitable that inhumane, unethical and immoral individuals will occasionally develop in much the same way that individuals will be born with physical genetic abnormalities. It is an inevitable consequence of the way that evolution proceeds. No human personality behaviour has yet been discovered that is entirely due to genes, so genetic make-up should not be used as an excuse to justify antisocial behaviour.

A priest once told me, as I was pondering my faith as a teenager, that God worked in mysterious ways when I asked him why suffering and evil existed. I didn't buy it, and as I read around evolution, I realized that it worked in understandable ways. Suffering and evil, although unpleasant, are an inevitable feature of our chance existence. My personal view is we should do what we can to help people who live with debilitating genetic mutations, but we shouldn't tie ourselves in knots trying to understand why these people exist. And of course, their lives can be as fulfilling as those of people who developed more typically.

It does appear that religion has moved on over the last few decades. When I asked a vicar friend whether an evil Neanderthal would have gone to hell, he said he didn't think hell existed, and he was even a bit vague on the definition of evil. He was also adamant that he didn't believe in a god of the gaps. If I am honest, I left being a little bewildered by what he did believe in, but he did acknowledge that some of his congregation did believe in a god of the gaps. It is a concept I struggle with.

The 'god of the gaps' argument states that science has made great progress but hasn't been able to answer all questions concerning our existence. I agree with the bit about science. There are some scientific hypotheses, such as the Many Worlds theory, that experiments may never be able to test. However, the gaps have been getting smaller. Deity advocates used to argue that something as complex as the human eye could never evolve, but through careful study of how the eye develops, and of more primitive eyes in other species, biologists have been able to describe the evolution of complex eyes from much simpler light-sensing cells. The human-eye-is-too-complex-to-have-evolved argument has been discredited, and one of god's gaps was bridged.

Current god-of-the-gaps advocates choose to focus on traits like the love of art, music or literature, or feelings of love and empathy, or even the feeling of being special. But it is a risky game. Once the ability to see, hear, smell, taste and touch evolved, life was able to avoid things that could cause damage while seeking out things that helped it thrive. In the presence of danger, life evolved flight-or-fight hormones such as adrenalin to help it escape, while chemicals associated with well-being such as endorphins attracted it to safe havens. It is these chemicals that make danger feel bad and safety feel good. These feelings are expressed in species with even a small degree of sentience, including shrimp and potentially even flies.

As our ancestors evolved to communicate more effectively, language and art developed too. *Homo erectus* may have had language, and Neanderthals certainly did, so our distant ancestors may have sat round fires at night describing where food had been found that day, or where dangerous animals had been seen, or how they had avoided death through deeds of derring-do, and the path was laid for story-telling. Adding a cadence to oft-told stories proved a great aide-memoire, and the first music was produced. Our ancestors, and Neanderthals, also created rock art. Recent research has shown how markings around pictures of animals represented a calendar, explaining when migratory beasts were present in the locality providing a source of food. Some of the earliest art had meaning, and being able to appreciate it would have offered a survival advantage, as it provides cues as to when to be in a particular area. Although an appreciation of Picasso or Rodin does not provide any individuals any evolutionary advantage these days (unless of course you met your partner in an art museum), the roots of our love of the arts may well have done.

As scientists research to fill in some of the gaps in our knowledge it is inevitable that there will be some speculation. That is how hypotheses are posed, and is an early step in any scientific endeavour to gain understanding. As we learn more about the ways our ancestors lived, and when language evolved, the existence of the phenotypic traits that are currently argued to be evidence of gaps will be bridged. That will not persuade many to stop believing in deities. Some people will always believe in gods, I won't change their minds, but that shouldn't lead them to dismiss science. There will always be individuals who feel special and that their existence on our planet has been given a purpose by a higher power. My advice to them is to not make that purpose trying to undermine science, as it is constructed on very strong bedrock and it has changed for ever our world and our understanding of why and how it came to be here.

That application of science has created an astonishing road map of the history of the first 13.77 billion years of the universe, and this road map is humanity's greatest achievement. There is still a lot to learn, particularly in the domains of biology and consciousness, with advances being published every day. Technological advances also mean we can conduct studies on scales only dreamed of by researchers of yesteryear. In mid-2022 the James Webb Space Telescope sent back photos of galaxies that formed 200 million years after the Big Bang. The light left these galaxies over 13.5 billion years ago and has been travelling at 299,792,458 metres per second since. Astronomers are gazing back at the very early universe, and the photographs from the world's most powerful space telescope will allow scientists to refine our understanding of the universe from the first days it became transparent.

At the other extreme, physicists are planning future particle colliders. One group proposes to construct a collider with a tunnel that is nearly four times longer than the 27-kilometre tunnel at the Large Hadron Collider that could create energies seven times greater than those produced by the LHC. Such energies would refine our understanding of the early universe and how matter formed and, in time, we may even detect properties of dark matter, and possibly dark energy. I suspect we are never going to be able to produce energies to create new universes, which is probably just as well, as where would we put them?

Observations of other star systems are also helping astronomers estimate how common Earth-like planets and gas giants are, and even what their atmospheres look like. The technology is still limited to studying star systems relatively close to Earth. At the time of writing, the most distant exoplanet is a gas giant called OGLE-2014-BLG-0124L, 13,000 light years away, less than half the distance to the centre of the galaxy. Although astronomers have yet to detect a planet with an atmosphere like that on Earth that would be a clear signature of extraterrestrial life, this lack of discovery is not surprising given the first exoplanet was discovered only in 1992, and only a tiny fraction of star systems have been examined. There are at least 100 billion stars in the Milky Way, yet scientists have looked for planets around less than 4,000, while the study of atmospheres around these planets has been conducted on many fewer.

The discovery of life on planets other than Earth would

arguably be science's greatest achievement. Nonetheless, discovering life is not straightforward. An oxygenated atmosphere would probably suggest life, but a paper published in 2021 reported a model that showed chemical processes independent of life could result in an oxygen-rich atmosphere on water worlds. We may need another way to detect life on other planets. A number of other remotely detectable chemical signatures of life have been proposed, but it is also possible that inorganic chemistry could also produce such molecules.

The discovery of phosphine in the atmosphere of Venus demonstrates the challenge of detecting life on other planets. Phosphine is a byproduct of anaerobic respiration of some organisms on Earth, and it is difficult to make without life. Small amounts can be created by volcanic activity and weathering of some rocks, but the amount found in Venus's high atmosphere in 2021 hinted at life. The scientists who made this discovery were honest about their findings, but some of the media got overexcited. Reanalysis of the data using methods that corrected for potential sources of error suggested that concentrations of phosphine were lower than originally reported, suggesting a volcanic origin may be possible after all. Although unlikely, we cannot rule out microorganisms living in Venus's atmosphere and are unlikely to be able to provide a definitive answer until we send spacecrafts to collect atmospheric samples. Indeed, the best way to detect life on other planets will be to visit them. We have sent people to the moon, and ground rovers to the moon and Mars, but the exploration of other planets is still in its infancy. Our best bet on finding life on other planets is discovering signs of either extant or extinct microorganisms on Mars, but we have only scratched the surface – literally – in our search. Proving the absence of life is potentially harder than proving its presence.

Closer to home, chemists are working to understand how the first replicators arose, and how they joined with membranes and metabolism to form the first cells. Such work, coupled with computer simulations of complex molecules, is revealing possible routes to how life began. In the coming decades there is a possibility, though perhaps only a small one, that we will create primitive life in the laboratory.

The science of the spread of life, and of the eventual rise of humans, and the development of civilizations is based on less evidence than most of the other scientific endeavours described in this book. Fossils of our ancestors are rare, and piecing together narratives based on partial skeletons or fragments of ancient DNA is challenging. Technologies that chemically analyse fossils and the surrounding rock and soil have helped date key events, but the scarcity of evidence means the field is subject to change. Human evolution, and early hominin evolution, is a fascinating subject, but it currently involves a lot of speculation. More important archaeological sites will be found, and each will help us understand how our ancestors lived in a little more detail. We are a remarkable species, and understanding of our technological and artistic history will continue to grow. Studies of Neanderthals and other now-extinct hominins will likely provide stronger evidence that two intelligent species have walked the Earth in the last few tens of thousands of years.

In the previous chapter, I considered why we turn out as we do. Understanding how patterns emerge at the level of the population is what scientists tend to do, but robustly explaining why a particular individual turned out as they did requires an understanding of developmental mechanisms that is currently beyond us. We do know that at the level of the population, people who smoke are, on average, more likely to die at an earlier age than those who do not smoke, but that does not mean we can predict when any individual will die. I have occasionally heard people say things like 'My Uncle Harry smoked all his adult life, and he lived well into his nineties, so smoking can't be that bad.' Uncle Harry got lucky. On average, smokers do have shorter lives, but that doesn't mean they will all die before all non-smokers. Much as we can say that, on average, smokers will die at a younger age than non-smokers, and we can also make statements that extremely traumatic events early in life can impact personalities, it does not mean that all individuals are impacted in the same way or to the same extent. It feels to me as if there were some key events in my life, but I may be simply imposing a subsequent narrative to account for my past behaviour and the decisions I made. Not all twenty-one-year-olds who thought they might die of malaria decided to write an overly ambitious book, so I cannot prove this is what inspired me to take the task on.

At the start of this book, I described a universe rerun experiment to explore whether our existence was inevitable or down to chance. No one other than me really cares whether I would exist in another universe, and you are the only person who will be fascinated by your existence. The question was a little narcissistic, but if you're still reading, it was a successful hook. The question I was really asking is whether the universe is deterministic or stochastic, and I have concluded it is stochastic. Our existence is down to luck, and that makes us special. Nonetheless, our feeling of being special is probably an evolutionary feature of all sentient life. Two questions naturally arise: is sentient life inevitable in all universes, and how common is it in ours? We know the answer to the first of these questions: if the four fundamental forces had different strengths, then protons, neutrons, atoms, stars and planets could not form, and life as we know it could not evolve. The answer to the second question has not been answered as we are yet to find intelligent life elsewhere. There are billions of galaxies and trillions of stars. Many of these stars will have orbiting planets, and some of these planets will be suitable for life. My conclusion is that life will have emerged on a proportion of these planets, and on some it will have thrived, while on others it will have been killed off by some cosmic catastrophe or other. On those where life persisted for long enough, life would have become complex and sentient beings will have arisen. Some of these sentient beings may have been able to harness energy to build civilizations, and in some of these perhaps an alien author has written a popular science book on their existence. I can't be certain that has happened, but if it has, they will have uncovered the same forces and principles that I have described. Science finds universal truths, and it will be those truths that unite intelligent life in the cosmos. The universe is sometimes strange, but science has shown it does not work in mysterious ways.

I finish with a plea: for you to embrace science if you have not done so already. Our universe is governed by rules, with science being humanity's invention to uncover them. The way that science coaxes these rules to show themselves is profound, beautiful and inventive, and through applying scientific approaches to a whole host of problems scientists have come a long way in explaining our existence. There are still many questions that science has not yet answered, and as our understanding increases, new questions will doubtless be posed. Some of the workings of physics, chemistry, earth sciences and biology are still shrouded in mystery, but the progress that has been made over the last three centuries is staggering, and in the coming years much of the mysterious will be explained. The peculiarity of our existence is humanity's ultimate puzzle to solve, and science has completed quite a bit of the jigsaw of what had to happen for us to be here, but there are many pieces that are yet to be put in place.

I am extraordinarily privileged to be a scientist whose work on investigating how evolution operates has contributed ever so slightly to our understanding of why we exist. I am sure I would have had an enjoyable life if I had become a builder, plumber, electrician, hairdresser or professional cricketer, but if I had my time again, I would still choose to be a scientist. Each day I gaze on our universe with a mix of shock and awe, and revel in being conscious and having an admittedly incomplete and in places blurred simulation playing in my mind as to how I came to be.

In writing this book I did have moments when my lack of faith was tested. I personally don't think we will ever know why there is something instead of nothing, and that is frustrating, and at times while researching the chapters on life the chasm between relatively simple replicating molecules and the life cycle of humans seemed almost too vast for natural processes to cross, even in 4 billion years. Rationality won the day, and I kept a belief in a deity at bay, but I did conclude that if a god did exist, it would be a scientist. I use 'it' because I do not wish to assign form or gender to a deity. It wouldn't care about me or what I did, but it would want to identify the set of rules required for a universe to produce beings like it. I don't say this intending to be derogatory or disrespectful of anyone's personal belief, and I do not wish to raise scientists to an elevated status. I say this because if I could, I would run the replicated universe experiment I detailed at the start of this book, and in creating new universes I would presumably become some sort of deity in those universes that evolved intelligent life. My desire to create universes comes from a thirst for knowledge and a desire to learn more. If a deity was omnipotent and understood everything, why bother creating a universe when the outcome would be known with certainty?

These words will doubtless be seen as controversial by some. I may be accused of creating a god in my own image. I wouldn't be the first to do this, but I am not doing it as I do not believe in a god. Believers may accuse me of elevating science to the level of religion, while some scientists may accuse me of lowering it to that level, but that is not my aim. The reason I don't believe in gods is I don't believe we require them to explain our existence. I know that some people believe in a god even though they don't accept one is needed to explain our universe. But I adhere to Occam's razor, preferring the simplest possible explanation. Why believe in something that created us when our existence doesn't require it?

I know that belief in gods is about faith rather than requirement, and even if I could entirely scientifically explain all the whys and hows of why and how we came to be, there would still be those who believe in higher powers. I can't, and I don't want to, change their minds. My aim in writing this book has not been to change anyone's mind. Instead, it is to explain and to celebrate the achievements of science. I find it rather frustrating that a book on existence must still touch upon non-scientific explanations for our being, but it would be odd for a book on existence not to at least briefly touch on creation myths.

Science is a human invention, but it is not one you need to take on trust. You, like me, can do scientific experiments to test hypotheses to help explain our existence. I argued earlier in the book that we are all scientists at heart, and if this book achieves anything I hope it is that you engage with your inner scientist and join the hordes of scientists who are using humanity's greatest invention, the scientific method, to answer the most profound question we can ask: why is it that we exist? Intelligent life did arise in our universe, but our existence required a big dose of luck. Enjoy being conscious, for it doesn't last for long. Three score years and ten is but a blink of an eye in the 13.77 billion years our universe has been around.

Acknowledgements

I have loved researching and writing this book. I have had a huge amount of support and help, both knowingly and unknowingly, over the years from an army of friends and colleagues, and I owe them all a huge debt of thanks. I cannot list them all, but below are those who have been particularly important. First, thanks to my family. In particular, thanks to my wife, Sonya Clegg. Without her support, encouragement and endless conversations about science and existence, this book could not have been written. Thanks, too, to my children, Sophie, Georgia and Luke. Perhaps one day they will read the book in its entirety. Or perhaps not. They have had to listen to me wittering on about science and existence all their lives.

When we got Woofler as a family pet, Sonya, Sophie, Georgia and Luke agreed to regularly walk him. They did for the first week or two. Thanks for not walking him much after that, and thanks to Woofler for needing walks. It was during these walks around the parks of Oxford and Brisbane that I composed much of the prose I wrote down on my return home. I could not have completed this book without being forced away from the keyboard to think.

The first full draft of this book was written during a sabbatical year in Australia, much of it spent at my father-in-law's farm. Thanks to farmer Col for putting up with the Oxford don sat writing on the veranda, and his friends Eve-Ann Springate and Robert Hudson, who I spent many evenings chatting with. Thanks for the beer and good humour you provided while we sat around the best stove on the planet. Thanks too to Angela, Alex, Ava and Stella for making my year in Australia so enjoyable.

Rebecca Carter, my agent, has been amazing from the day I met her. Thanks for believing in me and in the idea of the book, and trusting that I could deliver it. Thank you also for answering all my questions about publishing, and for finding such a great publisher. Without Rebecca this book would not have seen the light of day.

Penguin Michael Joseph has been fantastic throughout the whole process. Alan Samson has been a wonderful editor. He has provided insightful and thoughtful comments that have improved the text no end, making incomprehensible first drafts of chapters accessible. Alan's patience, good humour and fabulous editing have made writing this book enjoyable. Dan Bunyard has kept the project on track, moving it seamlessly through each stage of the publishing process. Sarah Day has been a keen-eyed and expert copy-editor and Sukhmani Bhakar has organized the illustrations with supreme efficiency.

Claiborne Hancock and Jessica Case at Pegasus Books bought the rights to publish in the US. Thank you for believing in the book, and for being so wonderfully efficient in designing the US release. Margaret Halton, Terry Wong and Rebecca Sandell at Pew Literary have made securing overseas publication rights very smooth, and I am extremely grateful for all their hard work.

Bryony Blades, Sonya Clegg, Richard Dawkins, Jane Hodgson, John Park, Quentin Painter, Ana Rivero, Rick Stockwell, Tabby Taberer, Tianqi Wang and Andrew Wood all provided insightful comments on drafts of the complete manuscript, while Phil Burrows, Sarah Hilton, Chris Summerfield, Anna

Vinton and Wai-Shun Lau have commented on the chapter in areas they research. Their advice and correction of errors are greatly appreciated. Any inaccuracies or misrepresentations that remain are entirely my fault.

I am grateful to numerous friends and colleagues who have helped me become a scientist and thrive. In particular Steve Albon, Charlie Canham, Tim Clutton-Brock, Mick Crawley, John Lawton, Steve Pacala and Josephine Pemberton were inspirational mentors during my graduate and postdoctoral studies. More recently, I have had wonderful and illuminating discussions in Trinidad and Yellowstone with Ron Bassar, Dan MacNulty, David Reznick, Doug Smith, Dan Stahler and Joe Travis. I have also collaborated a lot with my close friends Jean-Michel Gaillard, Pete Hudson and Shripad Tuljapurkar over many years, and I have spent countless enjoyable hours discussing numerous topics with them. Their views on science and the way it is done have helped mould my take on the universe and how it works.

Conversations with Anne Carlson, and then Amanda Niehaus, gave me the belief I could write this book, and I am extremely grateful to both of you for your encouragement. My friends Greg Devine, Djami Djeddour, Dominique Eza, Mike Furlong, Sally Gibbons, Nigel and Gen Griffiths, Sunetra Gupta, Richard Hobbs, Loeske Kruuk, Bob Montgomery, Ian Owens, Rick Paul and Sue Scull are yet to read the book, but have all helped keep me sane and happy at various points over the last thirty-five years. I have had conversations with each on the meaning of existence at some point. Some conversations were more memorable than others, but primarily due to the good food and wine rather than the content.

Lastly, thanks to my parents, Anne and Patrick, and my sister, Fiona. Without their encouragement and support over the last fifty-five years I could not have pursued a career as a scientist nor written this book. Sometimes my parents have not been entirely certain what it is I do for a living, so perhaps this book provides some answers. I have been doing quite a bit of thinking. I know they won't agree with all I have written, but I do hope this book makes them proud, and I look forward to discussing life, the universe and everything with them for many years to come.

Further Reading

Each chapter in the book describes what had to happen for us to exist. There is a lot that is not discussed. If any part of the book piques your interest and you wish to learn more, I provide details of a few of the very many texts I have read while researching this book. Each book goes into more detail than I was able to. Because popular science books tend to be on a specific science such as chemistry or physics, I do not group them by the chapter titles in this book.

History of Science

Grayling, A. C., *The Frontiers of Knowledge: What We Know about Science, History and the Mind – And How We Know It*, 2022, Penguin

Hossenfelder, Sabine, *Lost in Math: How Beauty Leads Physics Astray*, 2020, Basic Books

Kaku, Michio, *The God Equation: The Quest for a Theory of Everything*, 2022, Penguin

Robertson, Ritchie, *The Enlightenment: The Pursuit of Happiness 1680–1790*, 2022, Penguin

Waldrop, M. Mitchell, *Complexity: The Emerging Science at the Edge of Order and Chaos*, 1993, Pocket Books

Wootton, David, *The Invention of Science: A New History of the Scientific Revolution*, 2016, Penguin

Wulf, Andrea, *The Invention of Nature: The Adventures of Alexander von Humboldt, the Lost Hero of Science*, 2016, John Murray

Physics

Al-Khalili, Jim, *The World according to Physics*, 2020, Princeton University Press

Al-Khalili, Jim, *Quantum: A Guide for the Perplexed*, 2012, Weidenfeld & Nicolson

Carroll, Sean, *Something Deeply Hidden: Quantum Worlds and the Emergence of Spacetime*, 2021, Oneworld Publications

Galfard, Christophe, *The Universe in Your Hand: A Journey through Space, Time and Beyond*, 2016, Pan

Hooper, Dan, *At the Edge of Time: Exploring the Mysteries of Our Universe's First Seconds*, 2021, Princeton University Press

James, Tim, *Fundamental: How Quantum and Particle Physics Explain Absolutely Everything (Except Gravity)*, 2019, Robinson

Rovelli, Carlo, *Reality Is Not What It Seems: The Journey to Quantum Gravity*, 2017, Penguin

Rovelli, Carlo, *Helgoland: The Strange and Beautiful World of Quantum Physics*, 2022, Penguin

Strogatz, Steven, *Infinite Powers: How Calculus Reveals the Secrets of the Universe*, 2019, Mariner Books

Chemistry

BBC Radio, *In Their Element: How Chemistry Made the Modern World*, 2020, BBC Audio

James, Tim, *Elemental: How the Periodic Table Can Now Explain (Nearly) Everything*, 2018, Robinson

Lane, Nick, *Oxygen: The Molecule that Made the World*, 2016, Oxford University Press

Pross, Addy, *What Is Life? How Chemistry Becomes Biology*, 2016, Oxford University Press

Astronomy and Earth Science

Cohen, Andrew, with Brian Cox, *The Planets*, 2019, William Collins

Gribbin, John, *13.8: The Quest to Find the True Age of the Universe and the Theory of Everything*, 2016, Yale University Press

Hand, Kevin Peter, *Alien Oceans: The Search for Life in the Depths of Space*, 2020, Princeton University Press

James, Tim, *Astronomical: From Quarks to Quasars, the Science of Space at Its Strangest*, 2020, Robinson

Loeb, Avi, *Extraterrestrial: The First Sign of Intelligent Life beyond Earth*, 2021, Houghton Mifflin Harcourt

Emergence of Life

Deamer, David, *Assembling Life: How Can Life Begin on Earth and Other Habitable Planets?*, 2019, Oxford University Press

Kauffman, Stuart A., *A World beyond Physics: The Emergence and Evolution of Life*, 2019, Oxford University Press

Knoll, Andrew H., *Life on a Young Planet: The First Three Billion Years of Evolution on Earth*, 2015, Princeton University Press

Knoll, Andrew H., *A Brief History of Earth: Four Billion Years in Eight Chapters*, 2023, Mariner Books

Nurse, Paul, *What Is Life? Understand Biology in Five Steps*, 2020, David Fickling Books

Evolution

Al-Khalili, Jim, and Johnjoe McFadden, *Life on the Edge: The Coming of Age of Quantum Biology*, 2015, Black Swan

Dawkins, Richard, and Yan Wong, *The Ancestor's Tale: A Pilgrimage to the Dawn of Life*, 2017, Weidenfeld & Nicolson

Gee, Henry, *A (Very) Short History of Life on Earth: 4.6 Billion Years in 12 Chapters*, 2022, Picador

Halliday, Thomas, *Otherlands: A World in the Making*, 2023, Penguin

Losos, Jonathan B., *Improbable Destinies: Fate, Chance and the Future of Evolution*, 2018, Riverhead Books

Shubin, Neil, *Some Assembly Required: Decoding Four Billion Years of Life, from Ancient Fossils to DNA*, 2021, Oneworld Publications

Consciousness

Cobb, Matthew, *The Idea of the Brain: A History*, 2021, Profile Books

Godfrey-Smith, Peter, *Other Minds: The Octopus and the Evolution of Intelligent Life*, 2017, William Collins

Godfrey-Smith, Peter, *Metazoa: Animal Minds and the Birth of Consciousness*, 2021, William Collins

Hawkins, Jeff, *A Thousand Brains: A New Theory of Intelligence*, 2022, Basic Books

Peterson, Jordan B., *Maps of Meaning*, 1999, Routledge

Seth, Anil, *Being You: A New Science of Consciousness*, 2022, Faber and Faber

Human History

Frankopan, Peter, *The Earth Transformed: An Untold History*, 2023, Bloomsbury Publishing

Graeber, David, and David Wengrow, *The Dawn of Everything: A New History of Humanity*, 2022, Penguin

Harari, Yuval Noah, *Sapiens*, 2015, Vintage

Higham, Tom, *The World before Us: How Science Is Revealing a New Story of Our Human Origins*, 2021, Viking

Honigsbaum, Mark, *The Pandemic Century: 100 Years of Panic, Hysteria and Hubris*, 2020, W. H. Allen

Personality

Brotherton, Rob, *Suspicious Minds: Why We Believe Conspiracy Theories*, 2016, Bloomsbury Sigma

Christian, Brian, and Tom Griffiths, *Algorithms to Live By: The Computer Science of Human Decisions*, 2017, William Collins

Dunbar, Robin, *How Religion Evolved: And Why It Endures*, 2023, Pelican

Gray, John, *Seven Types of Atheism*, 2019, Penguin

Mitchell, Kevin J., *Innate: How the Wiring of Our Brains Shapes Who We Are*, 2020, Princeton University Press

Yeo, Giles, *Gene Eating: The Science of Obesity and the Truth about Diets*, 2020, Orion Spring

Image Credits

Index

Page numbers in **bold** refer to illustrations.

INDEX

  238–40, 397–8
atmospheric pressure 152
atmospheric zones 150–1, 157
atoms and atomic structure 62
  ancient Greek understanding
    101
  complexity 117
  formation of 6, 72, 90
  ground state 117–18
  orbitals 114–16
  properties 108–9
  quantum behaviour 105–12, **112**
  reactivity 117–22
  size 120–2
  solar system analogy 113
  splitting 16
  stability 117–18
  structure 64, 65–6, 113–15, **115**
  wave function 108–11
  wave-particle duality 106–12, **112**
ATP 185–7, 207, 209
ATP synthase 186–7
Australia 132–4, 172–3, 322
Australian Tourism Board 133
Australopithecine species 308–10,
  311, 313, 317, 318
Australopithecus afarensis 309
Australopithecus anamensis 308
autocatalysts 203
autocatalytic reactions 200–3, 205,
  207, 209
axial spin 153
axial tilt 153–4
aye-aye 254, 255
azidoazide azide 96, 100

bacteria 215, 244, 249, 398, 399
bad behaviour 368–9, 408, 409–10

balloon metaphor 89–90
bananas, antimatter production
  95–6
Barnard's Star 139
best friends 288
beta particles 72, 96
big bang 86
biological diversity 213
biological legacies 174–5
biologists 372
biology, randomness in 388
birds 219, 303
birth order, and personality
  351–2
bison 255–6
black holes 82, 137
body shape, change over lifetime
  235–7
Bolt, Usain 342
bonobos 307
Bosch, Carl 337
bosons 72, 75
Botox 99
botulinum toxin 99
bow and arrow 326–7
brain
  3D models 355
  age related changes 355–6
  cell numbers 274
  comparison with computers
    385, 388
  complexity 272, 295
  and consciousness 264, 265,
    270–85
  cost 319
  decision-making 285–7
  efficiency 285
  electrical activity 267–70, 277–8
  evolution of 293–6

433

brain – *cont'd.*
 frame of reference system
  282–3
 information flows 277–9
 lack of mechanistic
  understanding of 354
 Neanderthals 319
 neuron overlap 279–80
 personality traits and 353–5
 prediction mechanism 283–4
 primary role 272
 simulation of the world 270–2,
  281–4
 size increase 296, 319
brain development 234, 347
brain injuries, and personality traits
 354–5
brain scanners 267–70
brain structure 262, 272–80,
 275, 279
 brainstem 272
 Broca's area 312
 cerebellum 273
 cerebrum 273
 glial cells 273–4
 hippocampus 284
 neocortex 273, 280–4, 296, 311,
  315, 384–5
 neuron networks 274–80,
  279, 355
 neuron orientation 280
 neurons 273–5, **275**, 280, 283–4,
  295–6, 347, 355, 384–5
 synapses 275, 277–8, 353–4, 385
brain-imaging studies 355
breeding failure, causes of 226–7
Brexit 408
Broca's area 312
building blocks 69

Bungle Bungle mountain range,
 Western Australia 132–4
bureaucracy 298–300
burial practices 304, 317, 335

calcium-40 96
Cambrian explosion 247–9
*Candidatus Prometheoarchaeum
 syntrophicum strain* MK-D1
 215–18, 221
Canham, Charlie 142–3
canine distemper virus 52–3, 54–5
carbohydrates 239
carbon 65, 92
 atomic structure 122–3
 covalent bonds 119, 123
 ground state 123
 isotopes 123
 and life 122–3
carbon cycle 337
carbon dioxide 66, 151, 165–6
carbon dioxide emissions 251
carbonyl chloride 99
Castor star system 130
Çatalhöyük 20, 324–5, 334–5
catalysts 125
cations 118
cause and effect 371
cave art 303, 321, 327
Cavendish, Henry 102
cells
 brain 274
 clumping 245
 differentiation 242–4
 division 346–7
 endocytosis 217
 energy sources 207–8
 enlargement 239–40
 eukaryote 240

evolutionary biologists 212, 304
evolutionary success 214
exoplanets 140–1, 168, 394, 414
exosphere, the 151, 157
exothermic reactions 124
expansion 62, 63–4
experiences, and personality traits
    355–71
experiments 6–7, 25–8, 47, 263,
    421
explosives 94
exponential growth 202–3, 216
extraterrestrial life 414–16
eyes 271

Fa Hien cave, Sri Lanka 326
facts, rejecting 22
faith 405, 420
fastest object produced 61
fatty acids 206
feedback loops 179
fertilizers 337
fire use 297, 304, 325
first law of thermodynamics 376
fish 234, 249–50
fish farming 333
Fisher, Ronald 45
fishing 326–7
flight-or-fight hormones 412
fluorine, reactivity 121–2
fluoroantimonic acid 121
food
    acquiring 254–5
    detection 254
    digestion 255–6
    energy allocation 256–8
    utilization 255–8
    variety 255–6
food security 334

force carrier particles 67–8
force fields 67–8
fossil fuels 250–1, 337
fossils and fossilization 21–2,
    126–7, 234–5
    Cambrian explosion 247–8
    human evolution 307, 309, 310,
        313, 315–16, 322, 416
    soft tissue 235
frames of reference theory
    282–3
Franklin, Rosalind 3
Frankopan, Peter 337
free will 293, 380, 381–3
friends, and happiness 288–9
frozen peas, fear of 359
fundamental forces 64–7, 168, 372,
    394, 418
    emergence of 86–7
    force fields 67–8, 72–3
    interaction 67–73
    just right 390–1, 403
    and star formation 134–5
    strength 129
fundamental particles 61, 73–5, **74**
    behaviour 62
    decay 75
    early interactions 70
    first generation 73–4
    formation of 5
    mass 77
    as probability distributions 381
    randomness 381–2
    scale of interaction 62
    second generation 74–5
    stochastic behaviour 381–3
    third generation 75
fungi 292
future